IMAGES
of America

PALMETTO

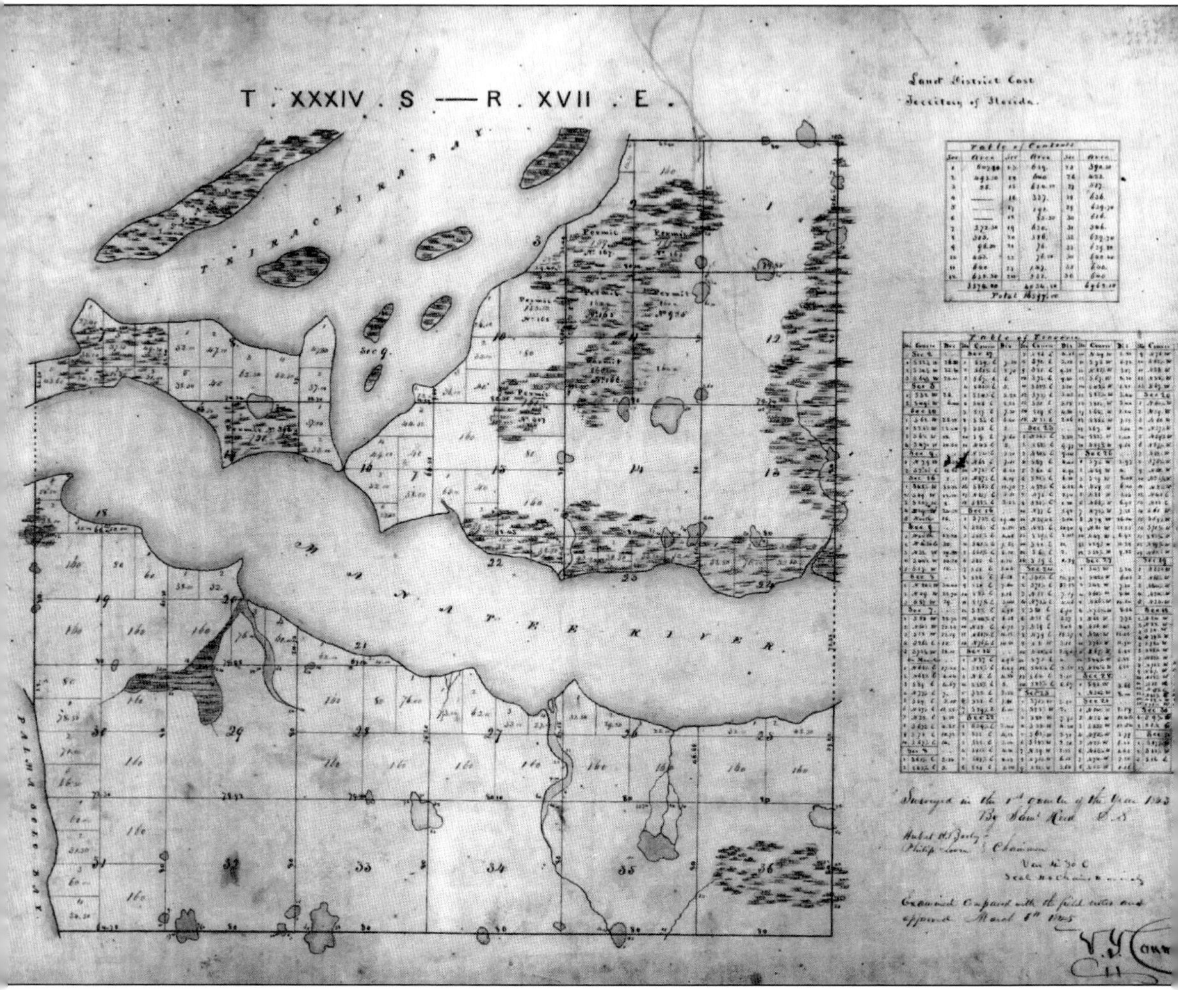

In this 1845 map, several plots are designated as pioneer properties. It was hand drawn by cartographer V.Y. Conray. (Courtesy of Neal Mazzei.)

On the Cover: This photograph by Harold R. Smith is an accurate depiction of life in Palmetto. This scene was captured in the 1940s or 1950s at the Seabreeze Park on Terra Ceia Island; however, a similar image could be captured at any one of Palmetto's present preserves where similar facilities are utilized daily. Note the fisherman concentrating on their catch while the women contentedly sun themselves. (PHC.)

IMAGES
of America

PALMETTO

Merab-Michal Favorite

ARCADIA
PUBLISHING

ISBN 978-0-7385-9104-9

Published by Arcadia Publishing
Charleston, South Carolina

Printed in the United States of America

Library of Congress Control Number: 2011931795

For all general information, please contact Arcadia Publishing:
Telephone 843-853-2070
Fax 843-853-0044
E-mail sales@arcadiapublishing.com
For customer service and orders:
Toll-Free 1-888-313-2665

Visit us on the Internet at www.arcadiapublishing.com

This book is dedicated to a fellow visionary and new friend, Jeff Burton, who took a chance and believed in me, and to my Journalism professor, Dr. Mike Killenberg, for never fully retiring.

Contents

ACKNOWLEDGMENTS

This book would not be possible without the extensive resources available at the Eaton Collection in the Manatee County Library, the genealogy/historic collection at the Palmetto Carnegie Library, the assemblage of African American history at the Family Heritage House Museum on the State College of Florida campus, the efficient employees of the Manatee County Historic Records Library, and the photographic collection of the State Library and Archives of Florida (SLAF). Through their dusty photographs, remarkable genealogies, worn paper articles, digital records, and biographies, this book was drafted.

Credit goes to the abundant archive of local interviews conducted by the Manatee County Historical Society (MCHS), by which this sleepy town's story unfolded and sprang to life. Special thanks to Pam Gibson, the library reference director and member of the historical society, who took time away from her busy schedule to help with historical dilemmas and records requests; also to Norma Dunwoody, who volunteers and devotes her time to preserving the area's little-known black history and helped immensely to uncover Palmetto's African American flavor. A very special thanks to Mandy Polson, Lynn Pope, Cindy Keck, and Diane Ingram, who allowed special access daily to sort through numerous collections at the Palmetto Carnegie Library. Their patience and kindness is the foundation on which this book is written.

Thanks to all the people who answered ads requesting family information, especially Lenore Stewart, James Allen, Nick Baden, Raymond and Beverly Varnadore, Tom and Evelyn Winter, all of the Groovers, Tony Haug, and Steve Pyles, whose generosity provided little-known treasures to this work. Thanks to Ken, Judy, Larry, and Nancy Burton for allowing interruptions at their Tuesday night family supper at Popi's Restaurant to listen to stories and verify fact from fiction.

Finally, to all the writers, journalists, historians, and photographers who over Palmetto's years recorded our priceless legacy, thank you, not just because of this book, but because this is my hometown. I will always support treasures like the Palmetto Historical Commission (PHC), whose primary focus is to keep history alive and well.

Introduction

In the mid-1800s, early settlers of the area encompassing present-day Palmetto puzzled over the mysterious shell mounds they found prevalent on both sides of the mouth of the Manatee River and on Terra Ceia Island. Did Native Americans construct them or had Spanish explorers erected them as a defense against adversaries? Clues to the origins of the mounds lay hidden until over a century later, when, in the 1950s, Ripley P. Bullen became the first archaeologist to study the Madira Bickel site on Terra Ceia. He found clear signs that treasure hunters and scavengers had long been at work digging and looting, including a particularly enterprising local man who hired a crew to bore holes to the bottom of the mound, hoping, in vain, to find riches. Despite the damage these interlopers caused, Bullen managed to uncover significant relics and human remains that help tell the story of Manatee County's earliest inhabitants.

Today it is widely acknowledged that native people constructed the mounds. Some mounds simply contained debris; others, such as the Madira Bickel Mound, were dedicated to ceremonial uses, including burials. To appreciate what these mounds represent for 21st-century residents of Palmetto requires a journey back to a time when herds of mastodon, giant bison, ground sloths, and saber-toothed tigers roamed the land.

Scholars believe Native Americans in Florida date back as far as 10,000 to 14,000 BCE. Generation after generation—for more than 3,500 years—the first Americans migrated and hunted prehistoric beasts. Gradually, the Paleo-Indian nomads in Florida settled into communities. They had little choice, for beginning around 8,000 BCE, rising global temperatures accelerated, melting away the final remnants of the Ice Age. As the climate changed, the oak and hardwood forests that had shaded part of the Florida peninsula morphed into more subtropical vegetation, like mangroves and palmettos. All the large mammals that once provided sustenance for native dwellers vanished; fossil evidence of their existence now lies beneath the coffee-colored waters of the Manatee River or concealed under the rich earth that today cultivates crops.

The great thaw also caused the sea level to rise dramatically, creating myriads of estuaries and bays teeming with shellfish and amphibians. Rising waters reduced the borders of prehistoric Florida by 200 miles. The new environment, however, with its extensive ecosystem, lent itself to a relatively easier existence for the Indian natives. Tribes began settling near the coast, surviving on the bounty of the sea and gathering rich vegetation that grew nearby. Around 6,500 BCE—referred to the Archaic Period by archaeologists—shell mounds, or middens, arose from the natives' cast-off refuse of clams, conchs, oysters, and scallops. Natives, armed with better weapons, lived off smaller game, like fox, opossum, raccoon, deer, and muskrat. They ate aquatic mammals such as dolphin and manatee. They consumed reptiles, including turtles, alligators, and snakes and gathered indigenous fruits, roots, and vegetables.

By 2000 BCE, the natives had adopted a semi-sedentary lifestyle. In the rainy seasons, they congregated atop the mounds where the breezes off the Manatee River and Terra Ceia Bay controlled mosquitoes and other invasive insects. To haul seafood and vegetables, they fashioned baskets from woven palm leaves and, later, created a crude form of pottery from sandy clay, tempered with fibers of Spanish moss or palmetto. The Neolithic Revolution had reached Florida; community life evolved and identifiable local cultures and dialects emerged. By 1400 BCE, villages engaged in widespread trade throughout the state.

With the influx of different cultures, burial practices changed. Instead of disposing of bodies in the wild or leaving them on the surface, natives began to intern their dead and adopt ceremonial practices. From 700 BCE to 500 CE, mounds specifically constructed for the dead, sometimes sanctified with charcoal, became commonplace. Bodies were arranged in the fetal position, and worldly belongings such as beads, pottery, and tools were placed on the east side of the mound and buried within. Some mounds played more hierarchical and religious roles. Many villages featured a temple—a pyramid structure with a flat top—where the chief priest, considered divinely guided, resided.

By 500 CE, farming had become the dominant way of life, but fishing (with nets woven of palm) and hunting game (with arrows and spears) still provided food for the natives. During roughly the same period, social stratification occurred, leading to a class system and accompanying inequalities among the native people. A chief priest assumed supreme authority over large territories. Special privileges marked status. For example, only people of high state were allowed to consume manatees; their meat was considered a delicacy.

When the Spanish arrived in the 1500s, the mound villages were well established, rising along the shore at an impressive height. The tribe that inhabited Manatee County was the Tocobagas. The Calusa tribe was to the south, and the Timacuan to the north. Archeologists have determined that the principal headquarters of the Tocobaga was either Snead Island or Safety Harbor.

Both men and women covered their bodies with tattoos of red dye and decorated themselves with shell jewelry and used fish bladders as earrings. The men wore deerskin loincloths, while the women made aprons of Spanish moss. With their hair styled in a high bun or a ponytail atop the head, they looked like giants to Spanish soldiers, although the tallest men were no more than five feet, eight inches tall.

The introduction of Europeans to Florida was swift and cruel. Spanish explorers searched every village looking for gold. The native population in the Manatee area was estimated at about 35,000. The armies required vast supplies, and soldiers were desperate enough to destroy anything that stood in their way. The natives were armed with bows and could shoot through several layers of chain mail, while Spaniards reported trouble even bending the tightly strung weapons. But the natives were no match for the soldiers and their powerful commanders. Both Hernando de Soto and Panfilo Narváez came through the area of present-day Palmetto.

Narváez landed at the mouth of Tampa Bay in 1528 with 400 men and 80 horses. At the village of Ucita, Narváez cut the nose off Chief Hirrihigua and had his mother thrown to the dogs. After hoisting flags of the Spanish crown, the army headed north, moving inland. Any Indian villages in its path became forging stations and the inhabitants subject to torture, slavery, or exposure to foreign disease.

In 1539, Hernando de Soto arrived with an armada of ships in the vicinity of the Manatee River. The fleet first anchored off Egmont Key, while a scouting party rowed ashore in small boats. The Spaniards made their first local contact of record at an abandoned Indian village—either Shaw's Point or at Snead Island. Later, de Soto brought the ships inland, scraping bottom on several occasions but pressing on, reaching a landing point on the south side of the Manatee River. After unloading livestock, provisions, and equipment, the larger ships set sail for Cuba, while the shallow-draft vessels were brought into the harbor. The army then crossed the river, also destined for Ucita, which is believed to have been located on present-day Terra Ceia. The Spaniards found the village empty, but they set up camp and stayed there for six weeks. When hostile natives surrounded the encampment, de Soto ordered the land cleared of brush and assigned guards to keep watch. By the time de Soto headed north, irreparable damage had already been done. Hundreds of natives died not from Spanish weaponry but from the diseases the soldiers unwittingly passed along to the natives. Modern-day archaeologists found evidence in the burial mounds of a variety of European-borne diseases. Children were particularly susceptible; large numbers of infants died.

While epidemics decimated the Tocobagas, the Calusa tribe from the south moved north to occupy the area. The Timacuan tribe of the north also may have lived around the Manatee River. During the late 1600s and early 1700s, researchers believe that Florida Indians traveled to Cuba on large dugout canoes to sell songbirds. The undertaking also brought in another wave of disease. By the 1760s, nearly all tribes native to Florida had virtually disappeared.

Besides a legacy of disease and death, the explorers left behind in Florida free-range cattle, horses, and hogs originally brought from Europe. The Seminole Indians in Alachua County became the first cow hunters, domesticating herds of cattle and riding Spanish ponies. But when the first Seminole War started in 1817, the tribe was pushed farther south, and Manatee County became a temporary sanctuary for Seminoles and runaway slaves. From the late 17th century, the Spanish crown had welcomed fugitive slaves to Florida, but with a catch—they must fight against the British to obtain freedom. During a brief period of British rule from 1763 to 1783, Seminole and Creek natives, as well as free blacks and escaped slaves, received British support to construct a fort on the Apalachicola River. Whites called it Negro Fort, and considered it a threat. In July 1816, American naval forces destroyed it, killing all but a few of its inhabitants. From the ordeal, the remaining Seminoles and blacks formed a union and integrated. A group of black Seminoles eventually took refuge along the Manatee River, calling their village Angola.

The ethnic and racial melding continued. Cuban anglers would spend the end of August through the end of March fishing the clear waters from camps and rancheros near the mouth of the Manatee River. They would salt and dry fish for the Havana market. The Cubans also traded in Angola for crops and skins, and soon they too integrated with the local population. When the United States acquired Florida from Spain in 1821, the government gradually pushed the descendants of Angola south, with some of them fleeing to the Bahamas. The Seminole fate in Florida was sealed after a tribe massacred 110 US soldiers traveling north from Key West. The US government made it a priority to capture and deport all Seminoles living in Florida.

Capt. William Bruce, a Hillsborough County resident, employed approximately 150 Native Americans and runaway slaves at a fishing ranchero he had set up at the mouth of the Manatee River. The US government frowned on the operation and destroyed it in 1840 as part of the elimination of Native Americans in Florida.

After years of military engagements with the Seminoles, the United States passed the Armed Occupation Act on August 4, 1842, intended to help drive them out. Under the act, the government would give any able-bodied man 21 years or older 160 acres of land anywhere south of Ocala, provided that he live there for five years, clear five acres of land, and build a house fit for human habitation. The plan actually worked, for it was an opportunity too good for people to refuse. Immigrants who owned nothing could find a place to call home in a location that seemed like a paradise. The settlers had no choice but to work together, forming neighborhoods and communities and protecting one another from the threat of dreaded Seminole raids that continued to linger in the humid air.

The peninsula that once witnessed the destruction of native culture and later provided an asylum for those escaping slavery and persecution was now a promising heartland of new beginnings. The settlers knew little of the life that lay before them.

One

The Pioneers of Palmetto

Families who took advantage of federal land grants under the Armed Occupation Act found it more difficult than anticipated to stay on their homesteads for the mandated five years. Small villages were susceptible to the scourges of Seminole raids and disease-borne illness. The settlers bartered for survival, and some storekeepers allowed farmers to pay for merchandise in crops.

Early pioneers either built pine-log cabins or wove huts from palmetto leaves. A shortage of nails meant that log structures had to be painstakingly mortised. Until the first sawmill opened in 1866, cut boards and windows were shipped in from New Orleans. Few could afford such goods, so early on settlers relied on wooden shutters to keep them secure. They slept on homemade mattresses stuffed with cured Spanish moss. Families kept their own plots of vegetables, fenced in to ward off foraging free-range cattle and hogs. But the hogs and cattle also fed the settlers, who often took turns hunting and sharing the meat with neighbors.

Despite the hardships, people continued to settle in the area of present-day Palmetto. In 1842, the collapse of the Union Bank in Tallahassee left a number of prominent families penniless. They trekked south hoping for a new start. Returning Civil War soldiers, lured by stories of Florida's fertile land, also brought their families southward. By 1870, the population had exploded on the outskirts of town. The residents were mainly white; only a few freed African Americans lived nearby, because when war with the North was imminent, plantation owners shipped their slaves to Louisiana.

In 1888, tragedy struck when yellow fever found its way to Palmetto, carried by a salesman aboard a ship from Tampa. Residents watched houses on the outskirts of town for the telltale sign of circling buzzards. At the time, no path led to Palmetto's Old Cemetery. As yellow fever took its toll, townspeople had to carve a trail wide enough to accommodate the passage of coffins. The outbreak ended, and by 1890 the town, soon to be incorporated, was comprised of 25 families. Several of them resided in their original palmetto huts.

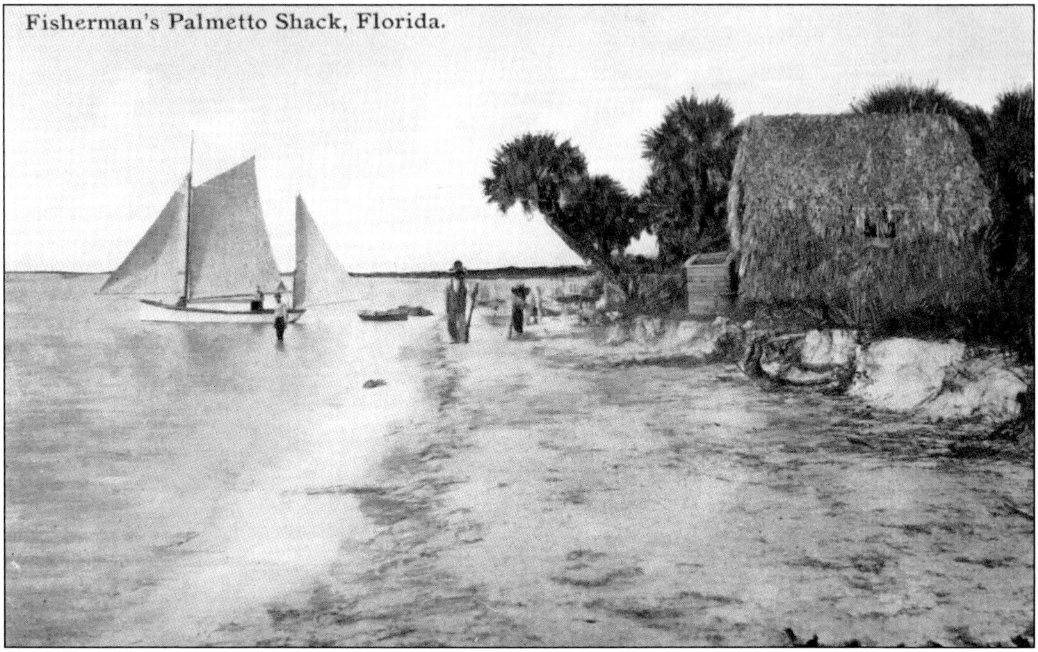

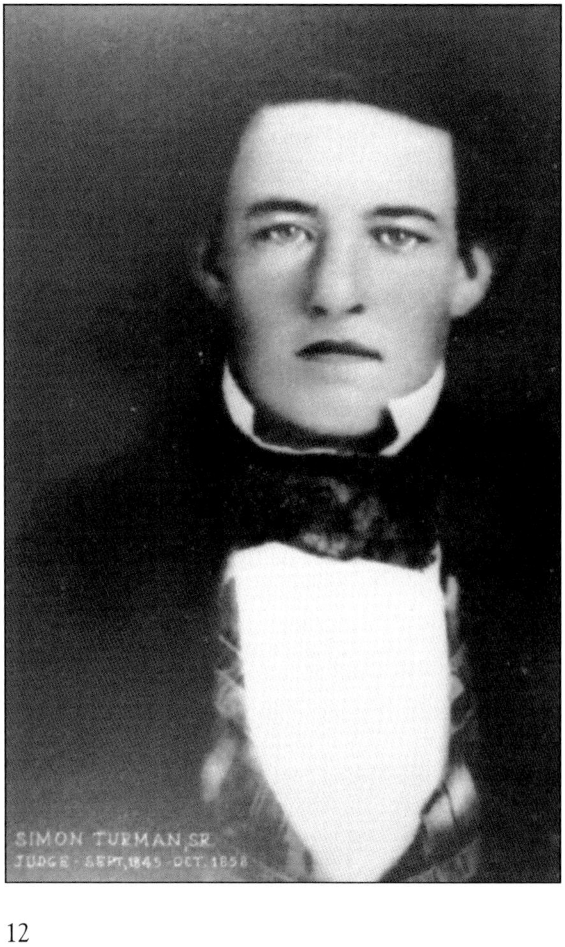

Seasonal Cuban fisherman constructed palm-frond huts near the shore as shelter. They dried and salted fish for trade in the Cuban market. The most popular species was mullet that they caught in nets made of cotton twine. Mullet roe was considered a delicacy. Today, Palmetto is still famous for the amount of mullet that congregate in coastal waters, and some locals continue to construct cast nets by hand. (SLAF.)

Simon Turman chartered a ship to Florida with five others in 1843. Turman acquired a mile of waterfront property that later flanked DeSoto Bridge. To receive ship traffic and imported goods, he constructed a cabbage-log crib in deep water called Turman's Landing. No dock connected it to the mainland, so cargo was rowed ashore. When the third Seminole War began in 1855, he planned strategies to cope with raids. (MCHS.)

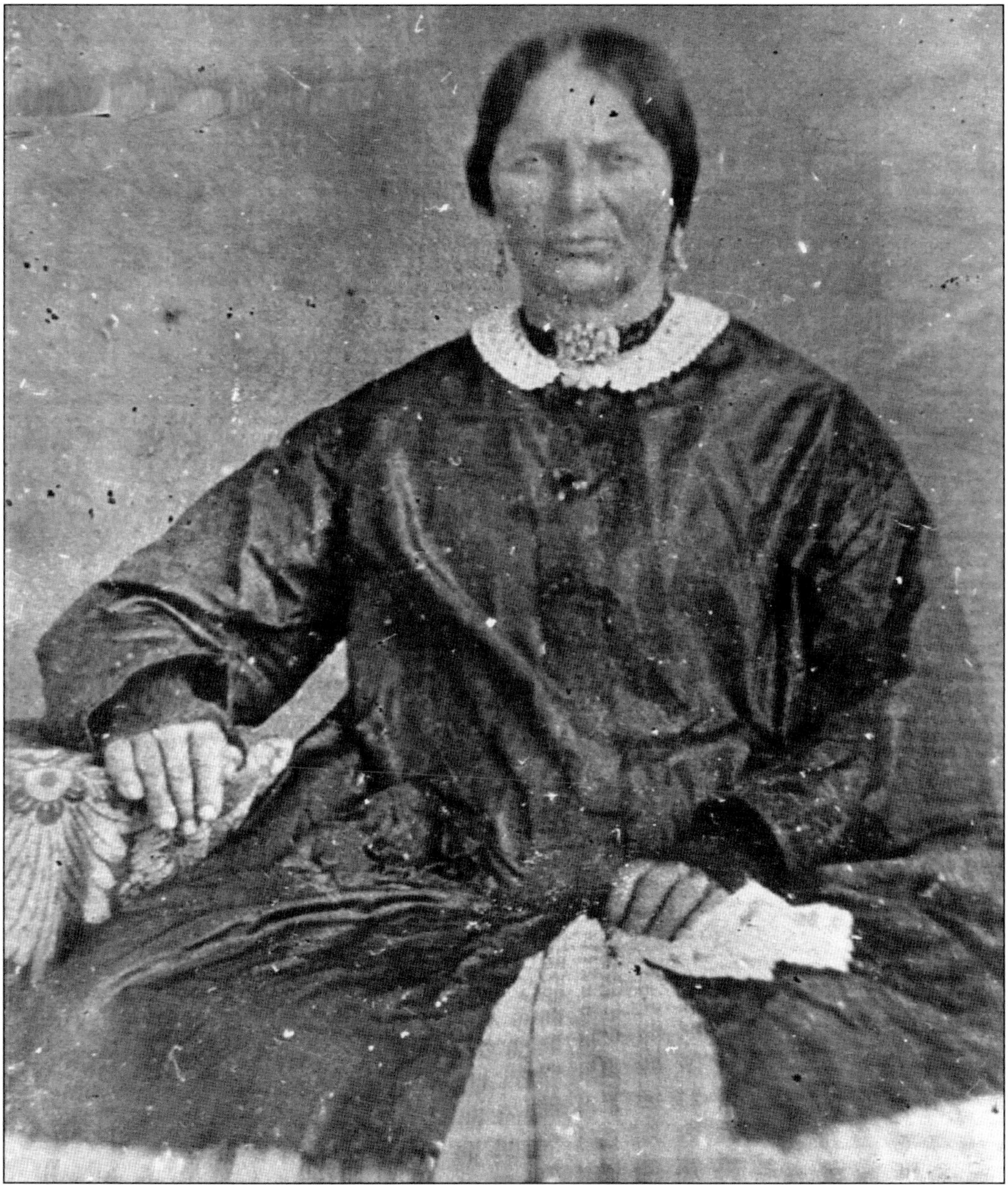

It seemed the only remedy needed for Julia Atzeroth's health problems was a grubbing hoe and elbow grease. Born in Bavaria, Germany, on December 27, 1807, she came to Terra Ceia in the spring of 1843 with her husband, Joseph Atzeroth, and acquired 160 acres under the Armed Occupation Act. When the government questioned the land title, Joe had to walk 160 miles to the land office to get the issue resolved. Because she always called her husband "Mr. Joe," she became "Madame Joe" to locals. An 1846 hurricane destroyed their home. The family lived in the only surviving structure, the hen house, until the home could be rebuilt. She ran a beer and cake shop in Tampa before moving to Palmetto, where she owned a general store with traveler accommodations. During the Seminole War in 1855, she stood guard with her musket whenever services were needed. After her husband's death, she moved to Fogartyville to live with her daughter, where she became the first person in the continental United States to grow coffee. (PHC.)

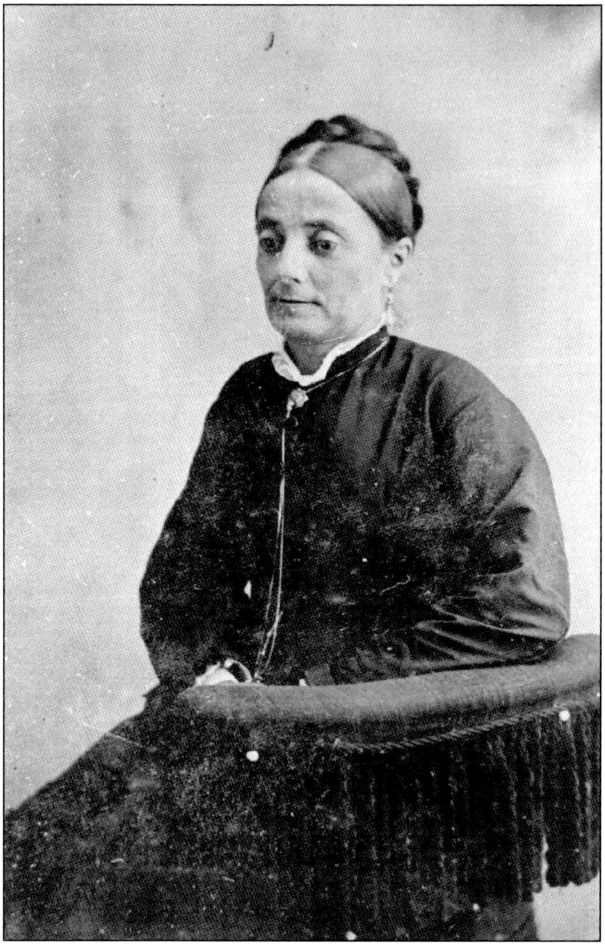

Eliza Atzeroth Fogarty, daughter of Joseph and Julia Atzeroth, arrived at age 11, becoming the fourth Caucasian child in the area. She married Michael Dickens at a double wedding ceremony with her cousin. Tragedy struck when Michael died in the Civil War and their two daughters died of disease. When Union soldiers boarded at her mother's inn, they tied a $5 bill around Eliza's neck as she slept because her mother refused compensation. She eventually married William Fogarty. (MCHS.)

Joseph Tillett (second from the left) and Wealthy Ann Webb Tillett (fifth from the left) sit on the front porch of their Terra Ceia home. Joseph, born in 1850, was a prominent fruit grower on the island for nearly 50 years. He helped build the Terra Ceia Methodist Church in 1899. Wealthy's father came to the area during the second Seminole War. She was born in 1849. (PHC.)

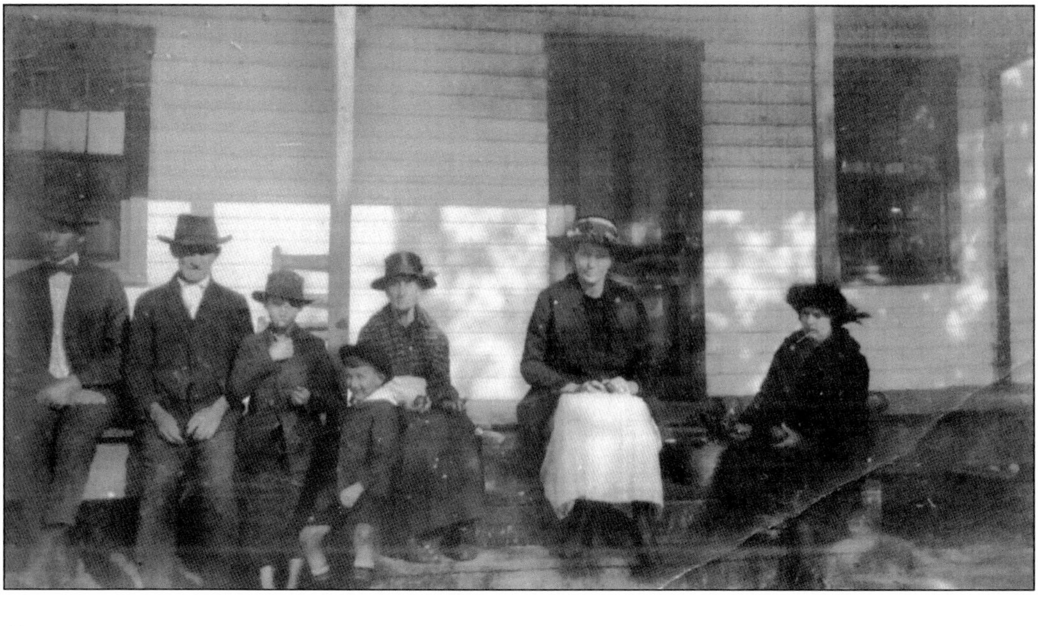

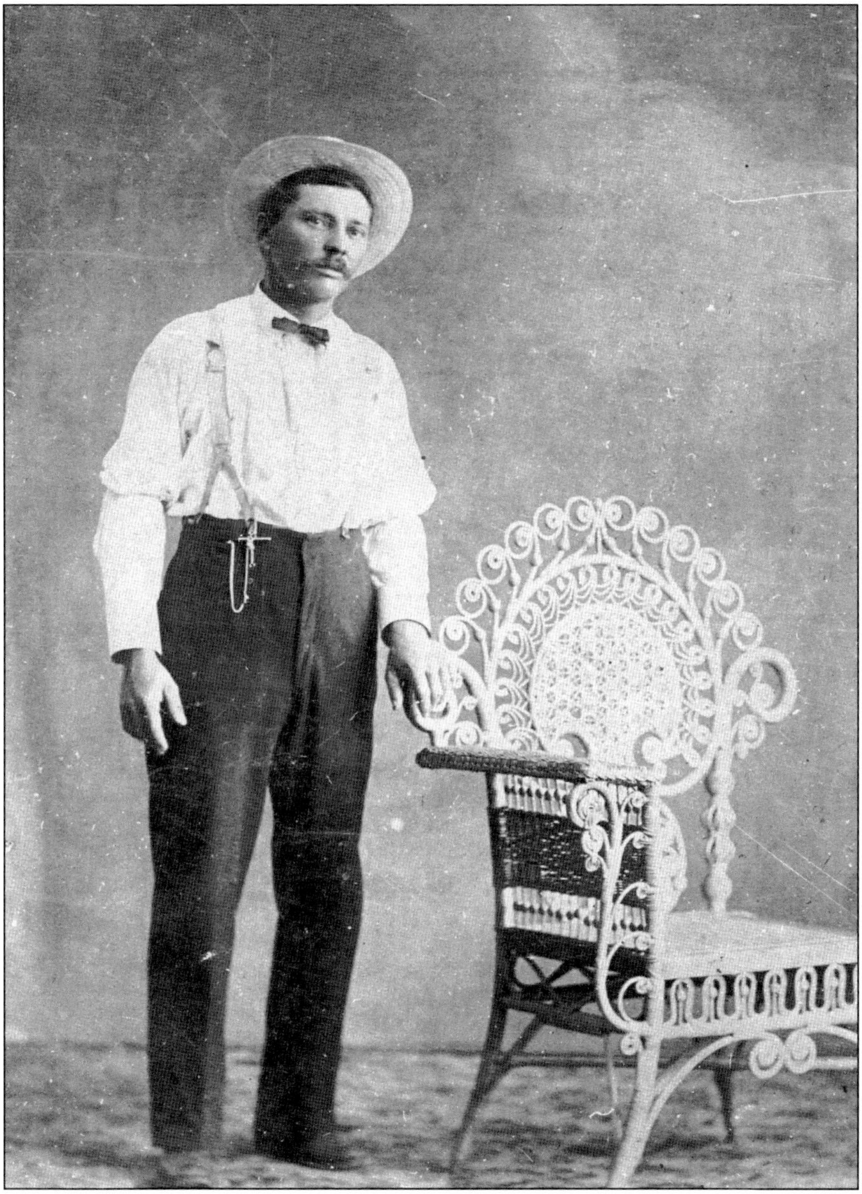

Robert Fogarty, born Robert Guerro in 1868, was found running wild with his siblings in the Terra Ceia wilderness. No one knew how long the children had been alone while their mother's body decomposed in the house. He was the son of Frederica Kramer Guerro of Bavaria, Germany, a niece of the Atzeroths. While visiting her relatives, Fredrica met and married Miguel Guerro, a Minorcan fisherman who homesteaded on the island. The two had five children, but a language barrier prevented them from communicating with one another. The children could not speak in either language and just uttered sounds that were discernable only to each other. While Miguel was away on an extended fishing trip, Frederica and two children died of yellow fever. Miguel also died of the disease, but the three youngest children survived and were rescued. Mary and John Fogarty adopted Robert, while the Lee family in Manatee adopted the other two. After extensive searching as an adult, Robert found his biological mother's grave on a creek embankment near the original home. He moved her remains to Palmetto Cemetery. (MCHS.)

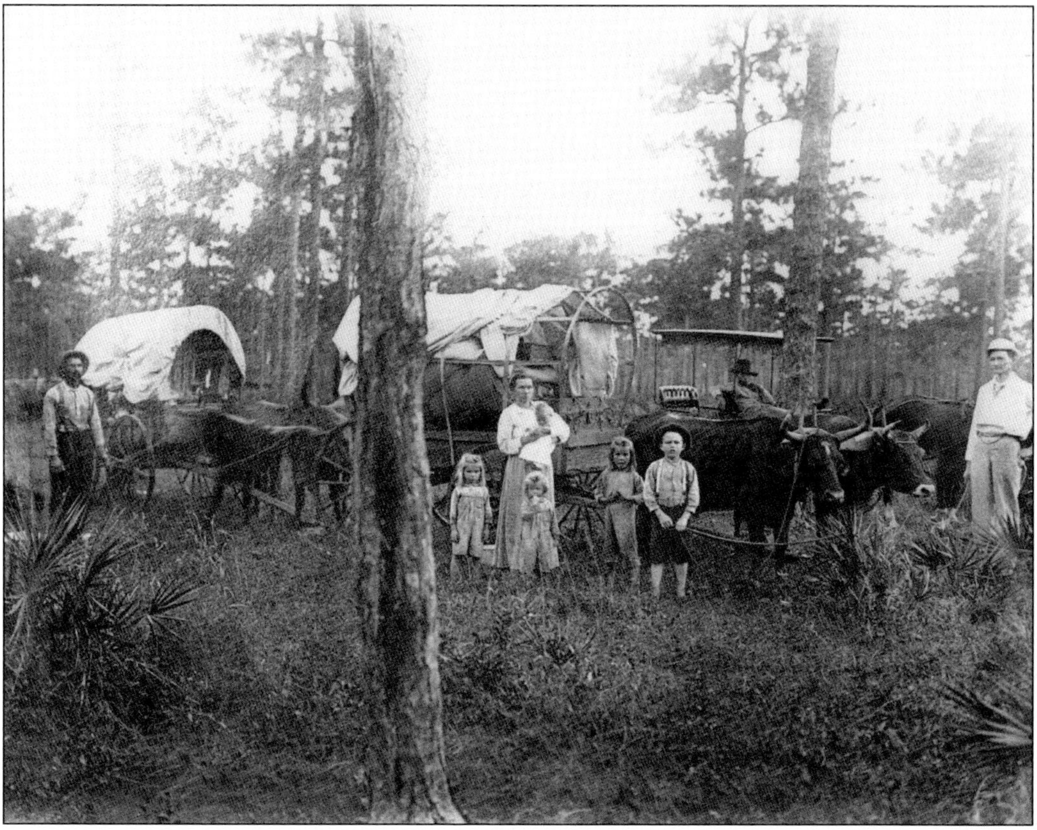

By 1846, county delegates requested a crude trail be cleared from Tampa to Turman's Landing. The blazed trail led wandering families to a village on the cusp of township. Caravans of travelers with loaded wagons pulled by oxen were a standard sight on the road to civilization. The trip took three days, making stopovers to hospitable houses a necessity. Unidentified settlers are seen here traveling on a Florida trail. (SLAF.)

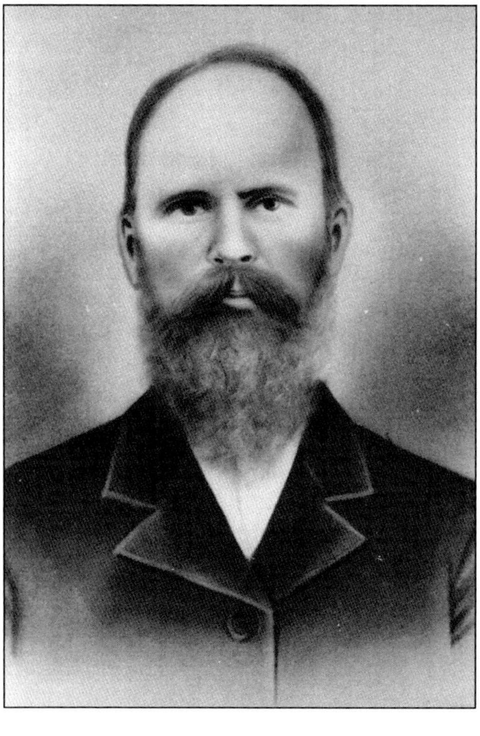

Daniel Uriah Richards was born in northern Florida in 1842. A group of hostile Native Americans killed his parents, but he and his siblings survived, remaining hidden in the woods and eating grass and leaves. Richards became a great cattleman and drove his herd to the area in late 1860s. He served as a sergeant in the Confederate Army and later married Mary E. Lamb. The couple had six children. (MCHS.)

Samuel Sparks Lamb (right) and his second wife, Sarah McLeod (below), ventured from Mississippi in 1868 when a doctor recommended they move south to cope with their son's health issues. Nine-year-old Julius "Dool" Lamb drove the barouche carrying his mother and five sisters while S.S. rode horseback, directing the oxen that pulled a supply-filled wagon. The family rested at Gamble Creek while S.S. enquired about land. He envisioned a town; he platted land and established Main Street (now Tenth Avenue) as a thoroughfare. He gave Palmetto its name because the plentiful palmettos reminded him of his birthplace, South Carolina. He is referred to as "the Father of Palmetto." Eventually establishing himself as Palmetto's most successful entrepreneur, he donated property for churches, a public library, and a park. He also set up the first school. (PHC.)

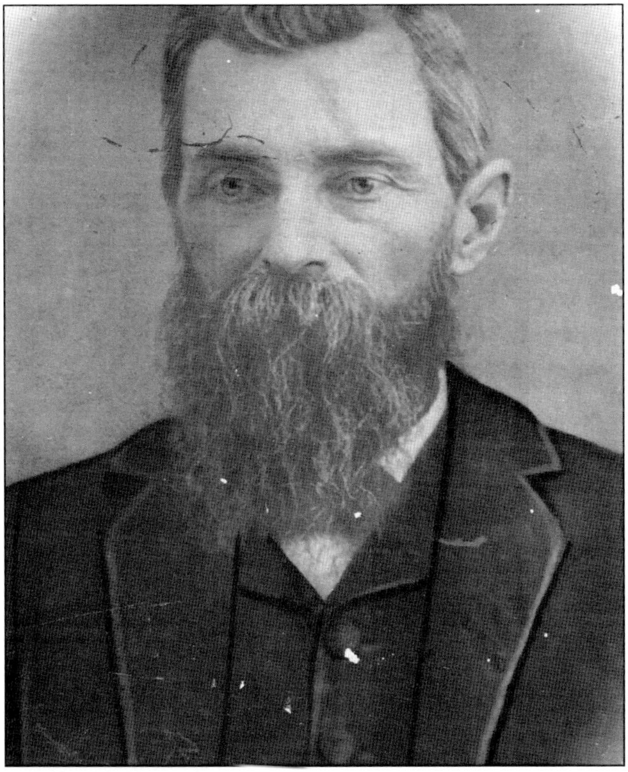

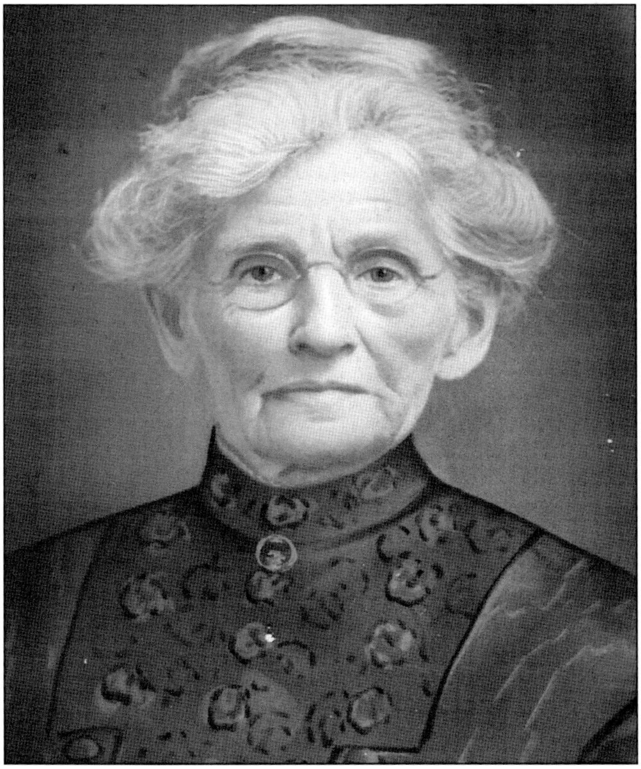

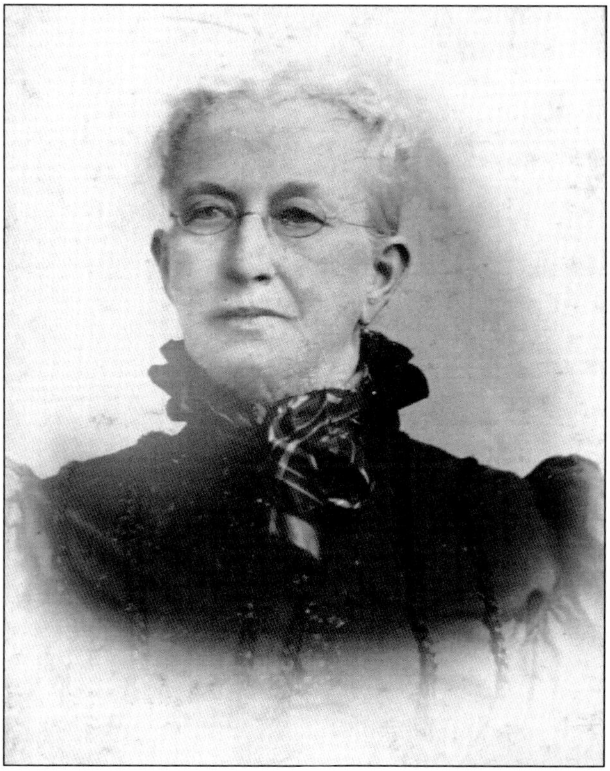

Sarah Hendrix Campbell (pictured), of Clarke County, Mississippi, purchased Madame Joe's land and outbuildings on November 27, 1866. She was the older sister of Joel Hendrix. She attracted other more permanent settlers. Soon after she homesteaded, her daughter in Mississippi became ill. She hastily sold her property to S.S. Lamb on February 3, 1868. (PHC.)

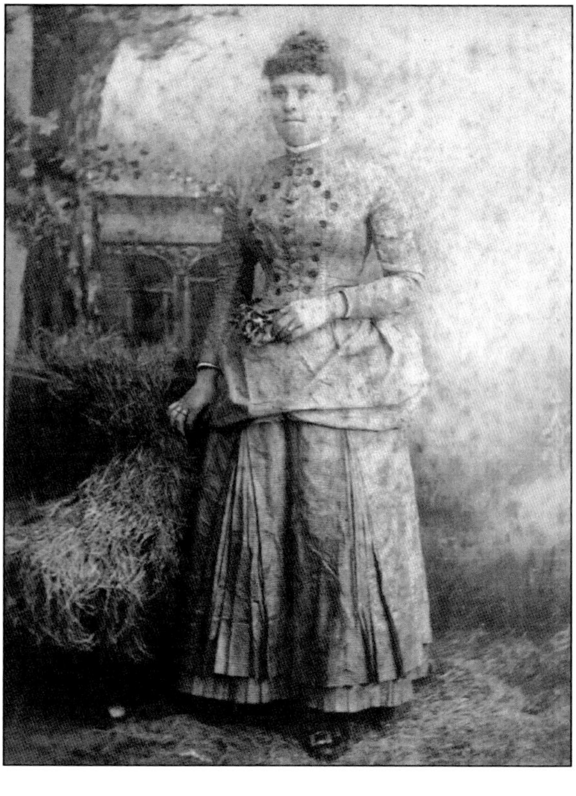

Born in May 1868, Laura "Lollie" Lamb Haley, daughter of S.S. Lamb and Sara McLeod, was the first person born within the Palmetto's official boundary. She married Charles Haley, and they had four children: Charles Jr., Julius, Sparks, and Katherine. (PHC.)

A hospitable African American couple, Mary and Joe Franklin, lived on the Snead Island Cutoff. When the tide went out, small boats were often stranded on the shallow stretch that separated the island from the mainland. The Franklins built two palmetto huts on the water: they lived in one and reserved the other for people held up overnight by the tide. They were members of the first white church in Palmetto. (SLAF.)

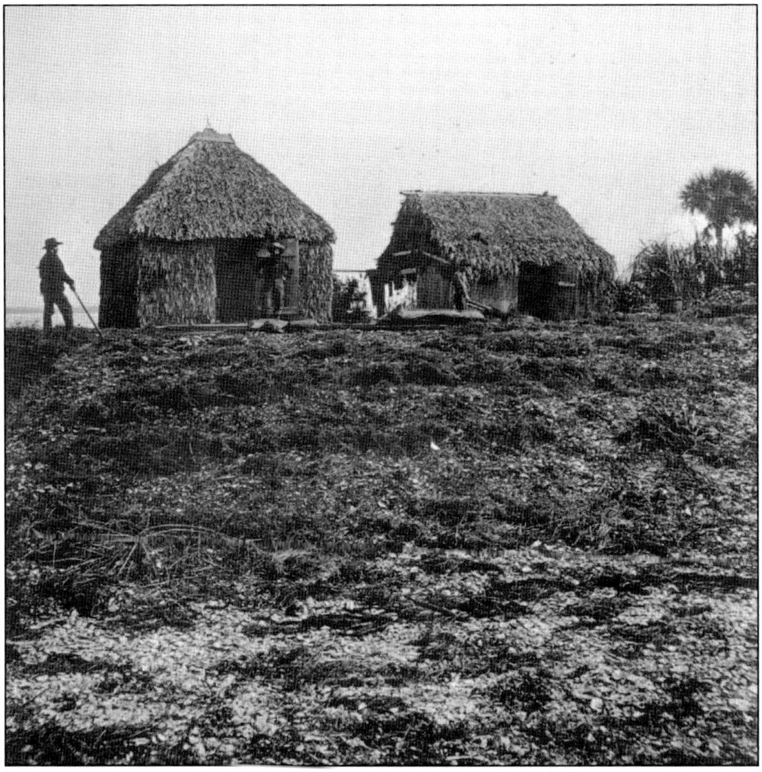

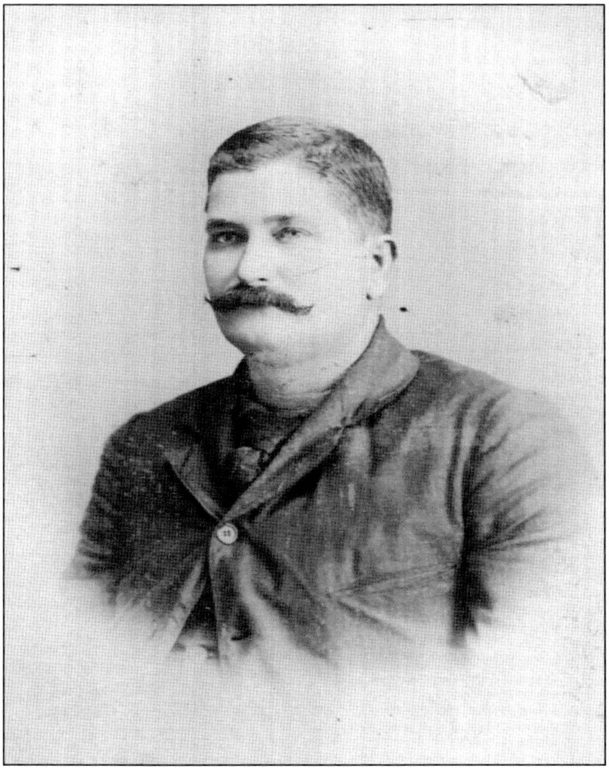

Joel Hendrix (pictured) and his wife Martha Ann arrived in 1871. Martha paid S.S. Lamb $100 for six acres of his property on the river. Joel built a home and general store on present-day Ninth Avenue and Third Street. One corner of the store became Palmetto's first post office. He constructed the first commercial wharf at the foot of Eighth Avenue after buying the property from his wife. (PHC.)

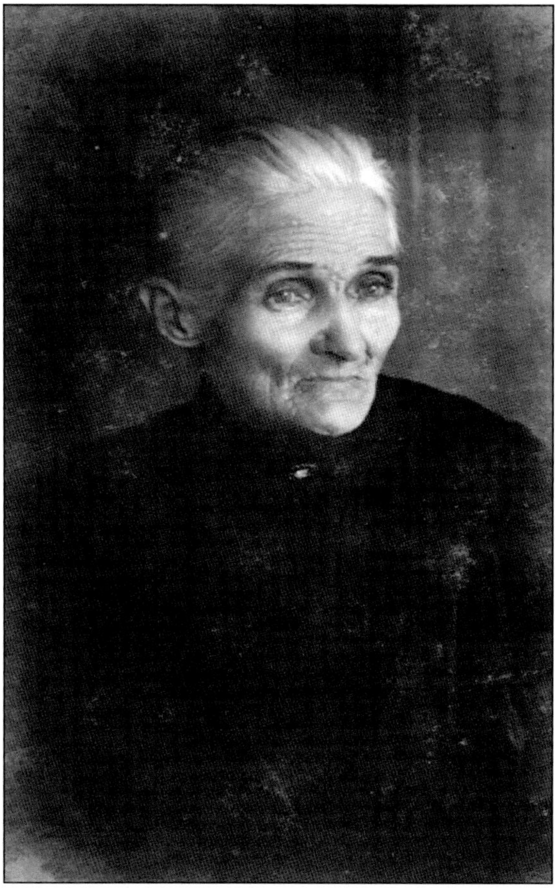

Pictured is Mary Nettles, wife of J.W. Nettles. She became the postmistress in 1880 after Joel Hendrix resigned the position. She initially operated the post office out of her home, but then she constructed the first building to be used specifically as a post office on Central Street (present-day Ninth Avenue) adjacent to her home. The building is now located at Palmetto Historical Park. (PHC.)

In 1882, Robert Willis (bottom left), who was the twin of John (bottom right), came from Memphis to join the rest of his family in Palmetto. Robert acquired land just north of Palmetto and platted the tract into lots, starting the community of Memphis, the area surrounding present-day Lincoln Middle School. Also pictured in the back row are brothers Virgil, Jim, Joe, and Haywood. (Courtesy Lenore Stewart.)

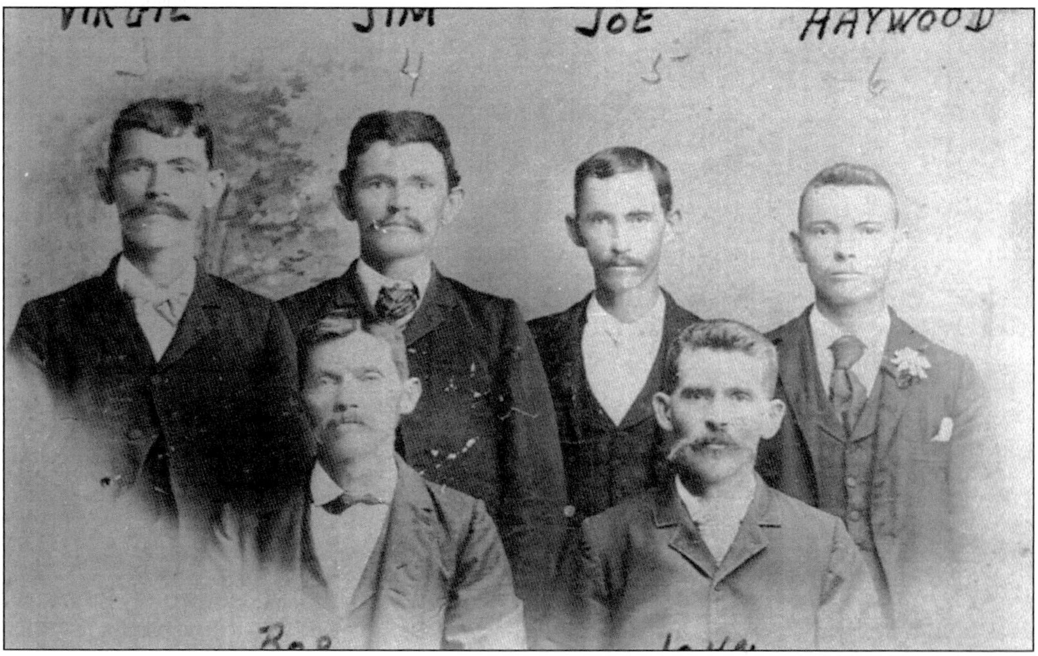

John P. Harllee Sr., Eloise E. Harllee, and John P. Harllee Jr. came from South Carolina in 1868 and purchased a mercantile store that sold farm supplies, including horses and mules. However, the animals were scarce in the area, so John's brother, Peter Harllee, traveled to Texas and brought back hundreds of horses and mules aboard a sailing ship. The venture proved to be very profitable for the businessmen. (PHC.)

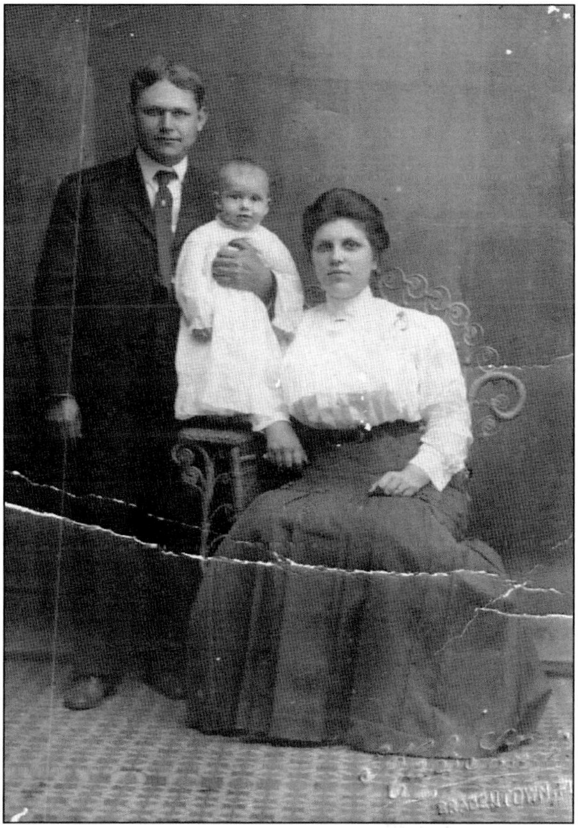

Peter Stuart Harllee, son of Peter and Alice Bullock Harllee, was born with yellow fever. His father moved the family to Palm View to escape the 1888 outbreak, but all six members became infected. The day after Alice contracted the fever, she gave birth to Peter Stuart. Miraculously, they all survived, and Peter Stuart became Palmetto's first mayor in 1894. He is pictured on the left posing with friends. (PHC.)

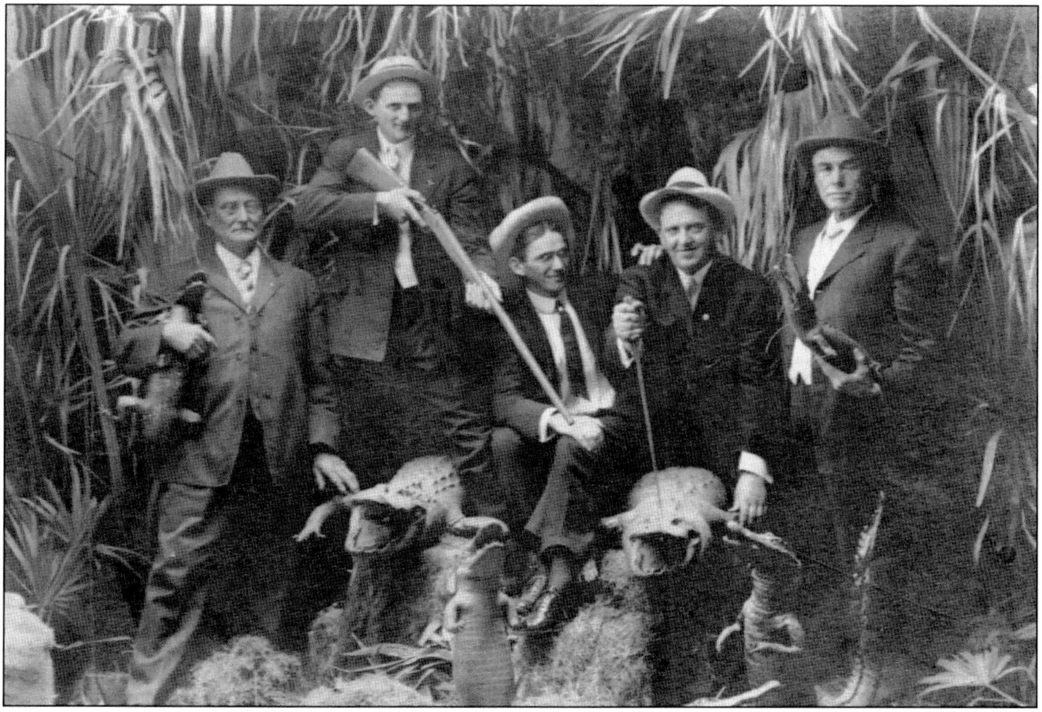

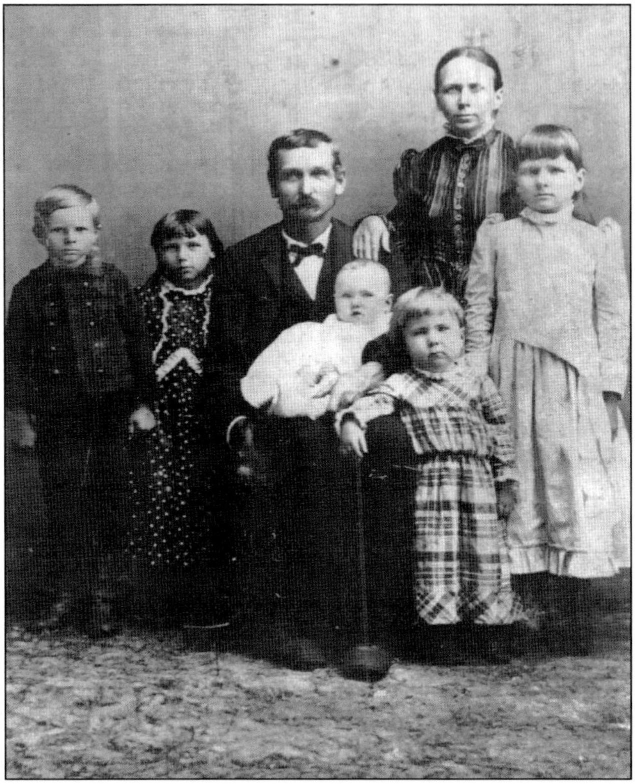

George E. Koons and his family pose for a formal portrait. The family resided on the northwest corner of Main and Seventh Streets in a palm-frond house with double-hung sash windows to keep the bugs out. Koons became postmaster on November 1, 1897, and served until March 2, 1915. (PHC.)

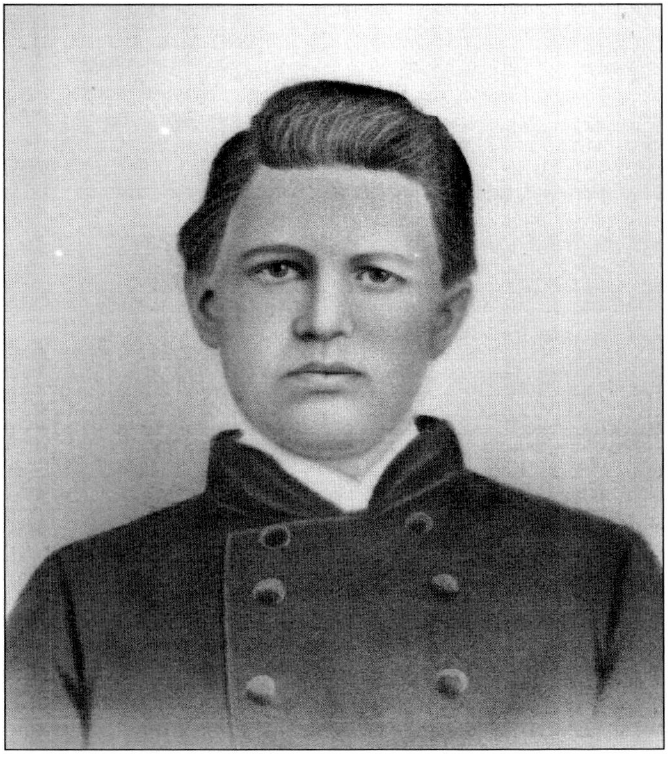

Capt. James A. Howze served in the Confederate Army. He and his wife opened a store west of S.S. Lamb's shortly after their arrival in 1888. Their entire family was the first to contract yellow fever. Mrs. Howze died, but James recovered and became a commercial dock owner, the first railroad owner, and a community leader. He was mayor when the city became incorporated and the first publicly bonded school was built. (PHC.)

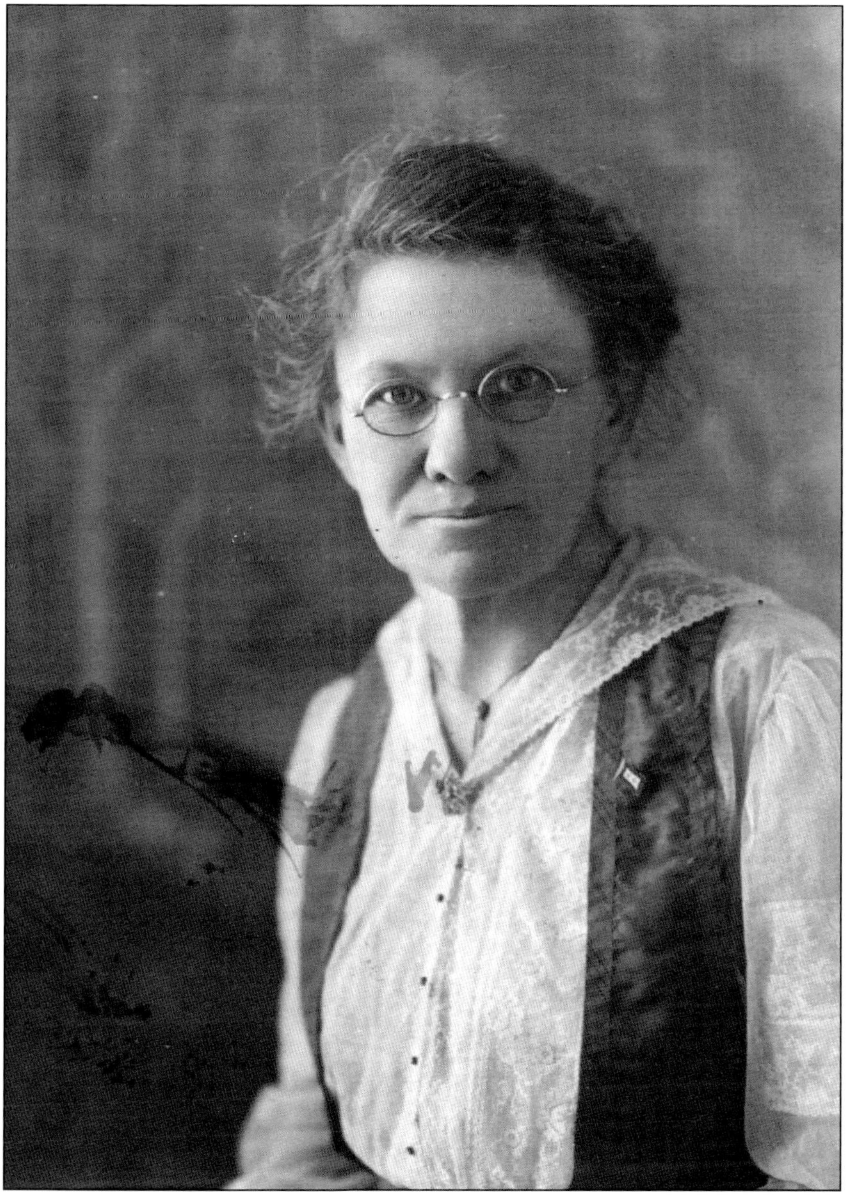

Francis (Frankie) A. McKay Howze is known as the mother of education in Palmetto. During her time as an educator, she was a teacher, a school principal, a librarian, and Palmetto's first football coach. Teachers who were revolutionizing methods of education in Chicago trained Frankie when she was a teenager. When her health failed, she traveled to Florida to live with a relative. In 1889, following the yellow fever epidemic, there was a need for teachers in Palmetto, and Miss Frankie answered that call. She ordered textbooks from Chicago and paid for them with her own money. She divided students into grades and became the school principal. As a vivacious young woman, she soon attracted and married Capt. James A. Howze, whose wife died of yellow fever. Frankie birthed three sons but was widowed when the youngest was nine years old. She taught school for 67 years, put her children through college, and took classes at Columbia University during the summer. Palmetto Elementary was named after her at one point, but after a restoration, the dedication was forgotten. (PHC).

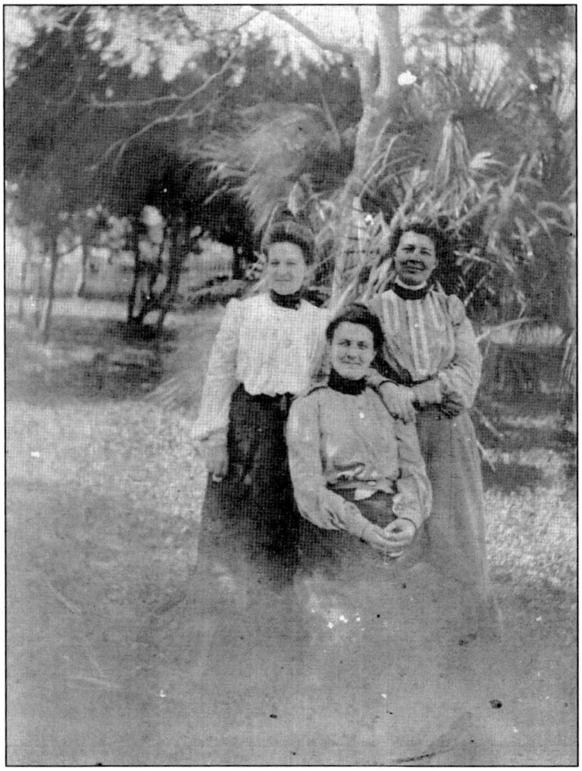

The first Mason sister to arrive in Palmetto was Nannie Mason McLeod (right) in 1894. She wrote her sister Lillie Mason (center), who came down on a steamer. Her beauty impressed Julius "Dool" Lamb, and they were soon married. Evie Mason Harrison (left) arrived later. (PHC.)

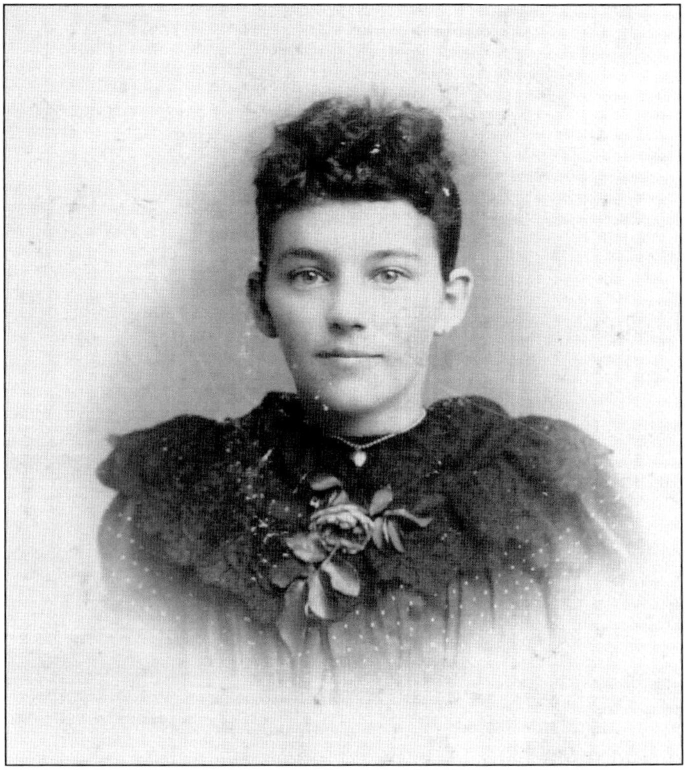

Nora Lee Willis was the daughter of John Willis. At six years old, she and her siblings walked a mile and a half from Memphis to Palmetto in order to attend Miss Frankie's classes at Palmetto's first school. This portrait of Nora Lee was taken in 1896 prior to her marriage to S.S. Lamb Jr. She is wearing a locket he gave her. (PHC.)

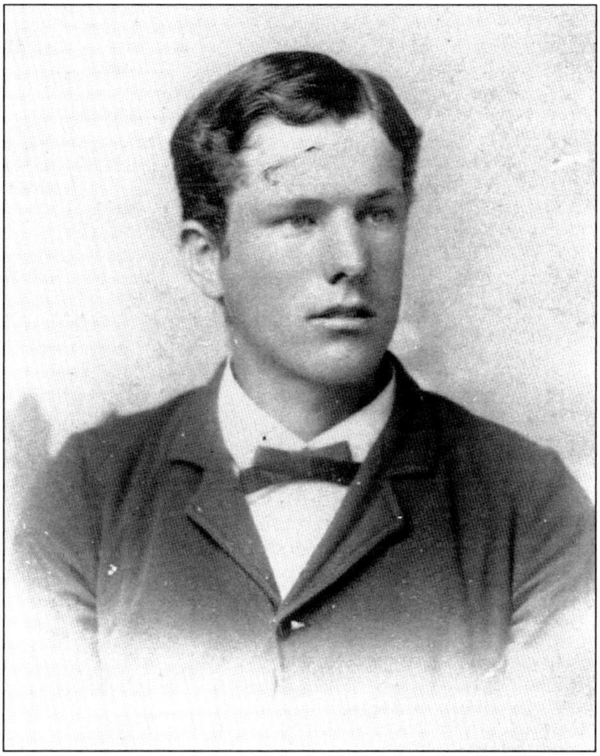

Capt. Edward Ithamar Pillsbury was born on September 24, 1895. He married Lorena Courter and had seven children. He started Pillsbury Boatworks on Snead Island, which was one of the most profitable boatyards in Florida. He used a washhouse next to his home as the first school on Snead Island. He was the government-appointed light keeper for the navigation aids at the mouth of the Manatee River. (MCHS.)

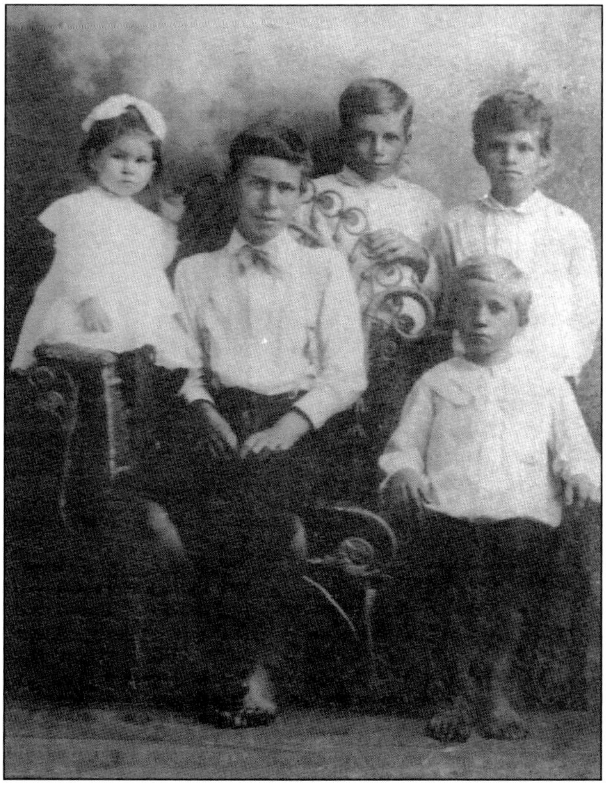

Aaron "Roy" Leroy, Elmer Levi, Asa Harmon, Edward Ithamar, and Ruth Iva were the children of Ed Pillsbury and Lorena Courter. Ruth was the first white child to be born on Snead Island in 1907. Their father floated the family home across the river from Fogartyville to present-day Snead Island Boatworks. The children never wore shoes, and the soles of their feet were so thick that they could walk across oyster beds. (PHC.)

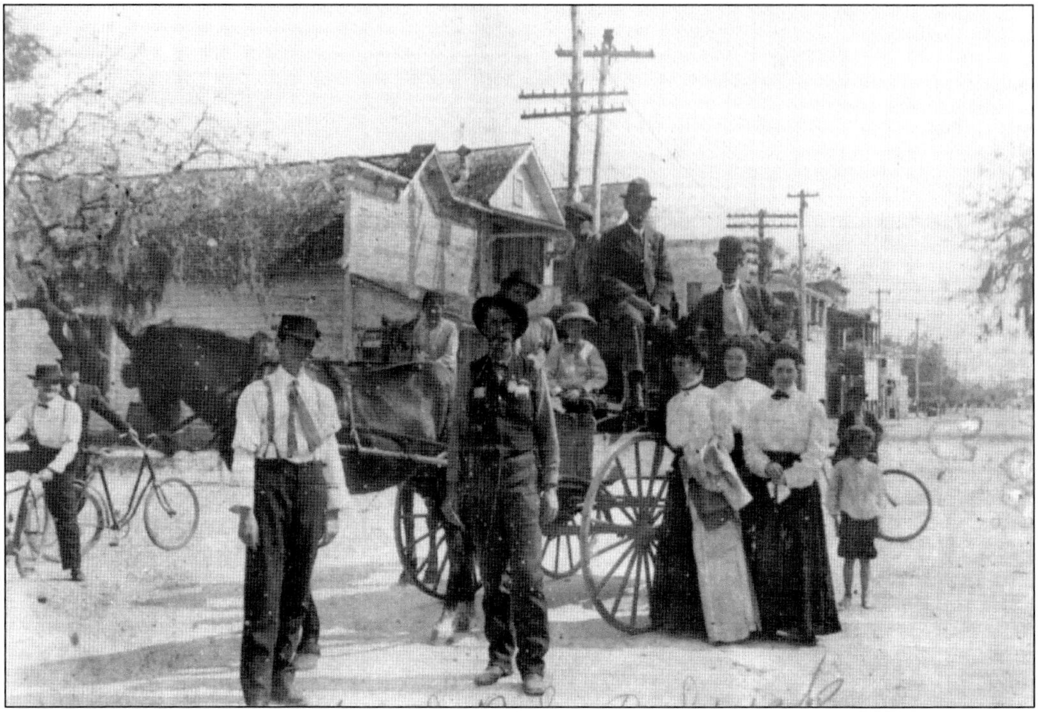

Town marshal Larry T. Pedrick and his family gather on Main Street in Palmetto around 1910. The Pedricks were one of the first successful farming families in the area. From the look of their attire, they were probably on their way home from church. (PHC.)

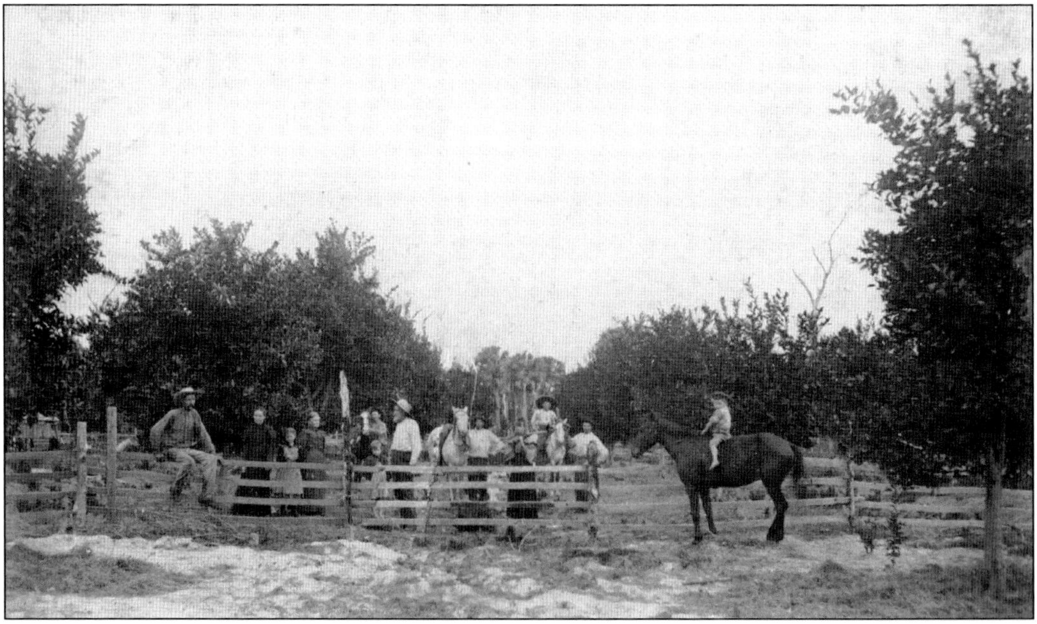

The Jackson family settled in Palm View. Included in this image are Mr. and Mrs. Nicols, a Nicols child, Margaret Jackson, and William Decator Jackson. (Courtesy James Allen.)

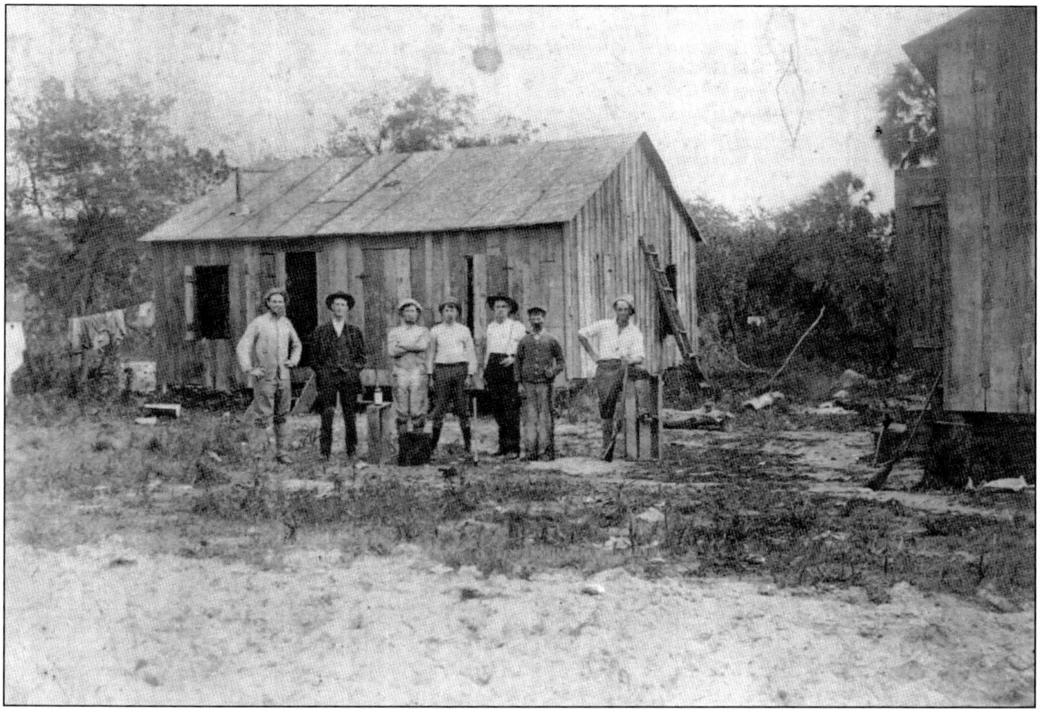

The Moon family settled near Frog Creek. Moon family members are seen here posing for a 1913 photograph. An inscription above the photograph reads, "The Bachelor's Hall." The buildings in the background are standard pine-plank outbuildings without glass windows. Shutters were used to keep bugs out but prevented air from flowing, causing the interior to be stifling in the Florida heat. (PHC.)

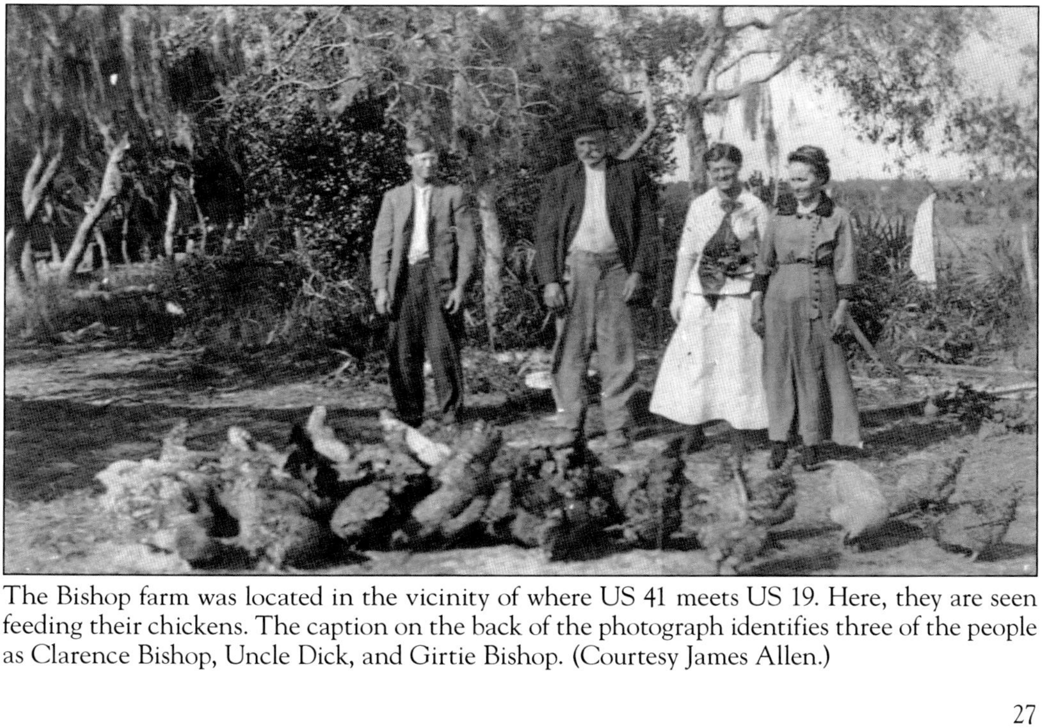

The Bishop farm was located in the vicinity of where US 41 meets US 19. Here, they are seen feeding their chickens. The caption on the back of the photograph identifies three of the people as Clarence Bishop, Uncle Dick, and Girtie Bishop. (Courtesy James Allen.)

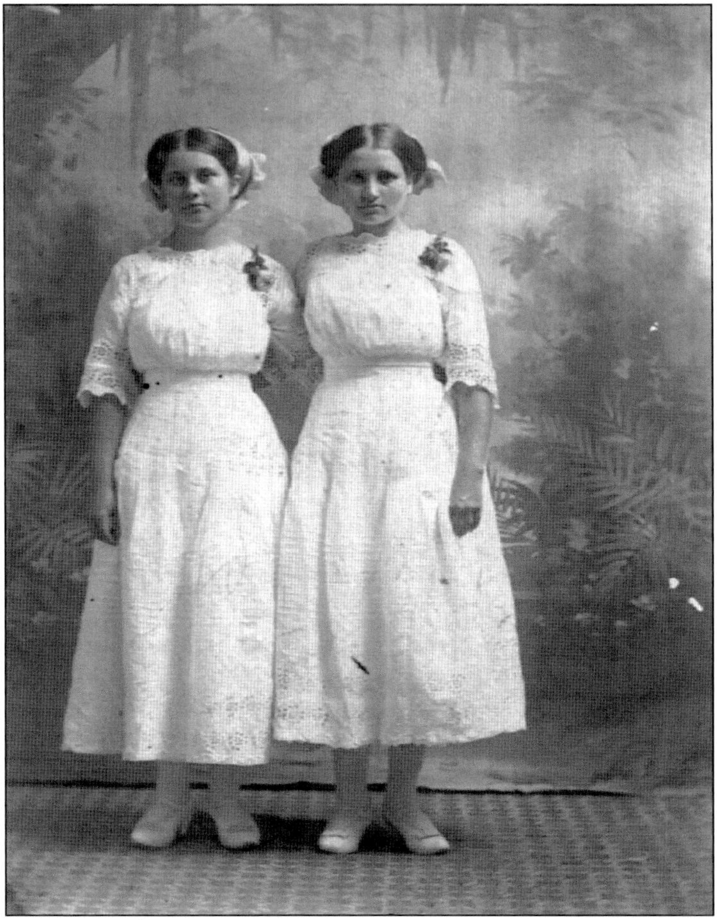

Mollie and Wealthy Rowell were the daughters of John C. Rowell and Julia Tillett. Their father was a prominent businessman and sent the girls to business school at a university in Tampa. Wealthy married John E. Anderson, and they raised eight children on Terra Ceia. Their original home is still located on Center Road. Mollie married Walter Lowrimore but became sick and died. Wealthy raised their son Trevor. (PHC.)

Pictured is the residence of Walter Eugene and Nellie Parris Lamb. Nellie is seen leaning on the fence in front; the man in the horse-drawn carriage is Dr. John Helms, who is making a house call. The house is a typical style of upper-class residents during that period. (PHC.)

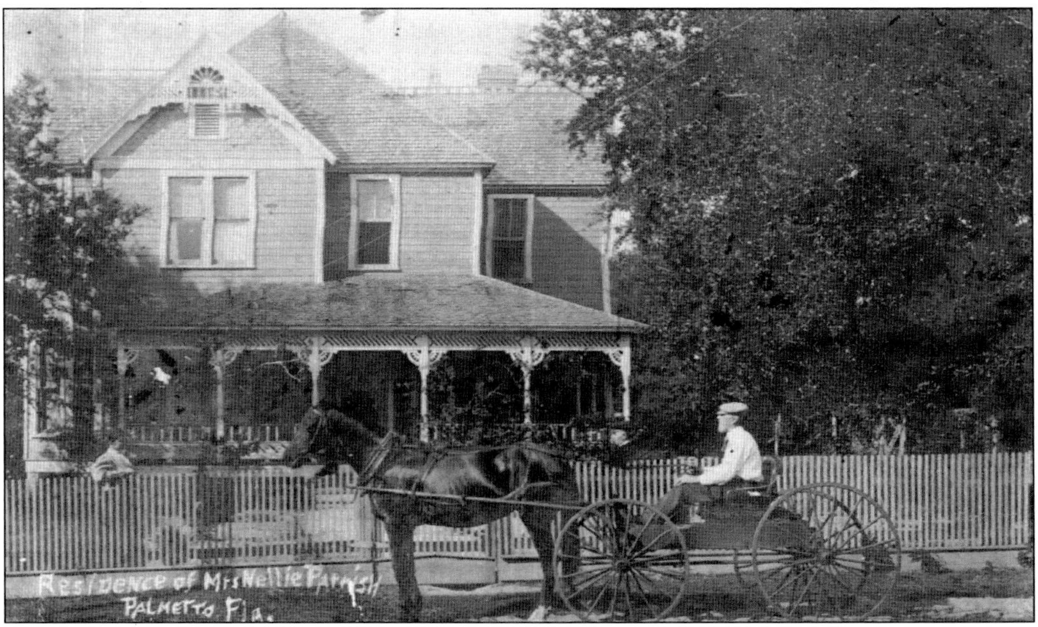

Two

Transportation by Sail, Rail, and Road

In the early days, there were no railroads or roads suitable for wagon traffic in the vicinity of Palmetto, and all transportation was by boat. The closest railroad reaching the west coast of Florida ran from Sanford to Cedar Key. From there, cargo and passengers headed south had to travel by sailboat. The fare was $8 for the 100-mile trip.

The earliest account of water transportation along the Manatee River details the role of a two-mast schooner named the *Margaret Ann*, captained by Fredrick Tresca. Archibald McNeil owned several schooners that hauled sugar from both the Gamble Mill, in today's Ellenton, and Braden Mill, near present-day Bradenton. Sugar was loaded at the Gamble Dock at Fourteenth Avenue East in Palmetto, bound for Baltimore, New York, Mobile, and New Orleans. McNeil's line of schooners operated from 1847 until 1861.

Manatee County was first part of Hillsborough County, and in 1846 Manatee citizens asked the Hillsborough County Commission for a road, which in those days meant cutting down trees and palmettos and removing stumps. The road meandered from Turman's Landing to Tampa, near the Alafia River. The trip usually took at least two days by wagon.

Steamers eventually replaced sailing vessels, and the bustle of river trade and travel enlivened residents. Steamships provided entertainment for families to enjoy a free ride up the river or an exciting day trip to Tampa for a bargain price. The first steamship to provide regularly scheduled service was the *Lizzy Henderson*, a 400-ton, wooden-screw steamer that made trips twice a month to Cedar Key, Manatee, and her home port of Tampa. Captain Gunderson operated the ship until the 1880s.

In time, railroads and paved roads replaced the river routes as a more efficient means of transport for cargo and passengers. The center of town, once so dependent on the river, shifted from Main Street to Eighth Avenue.

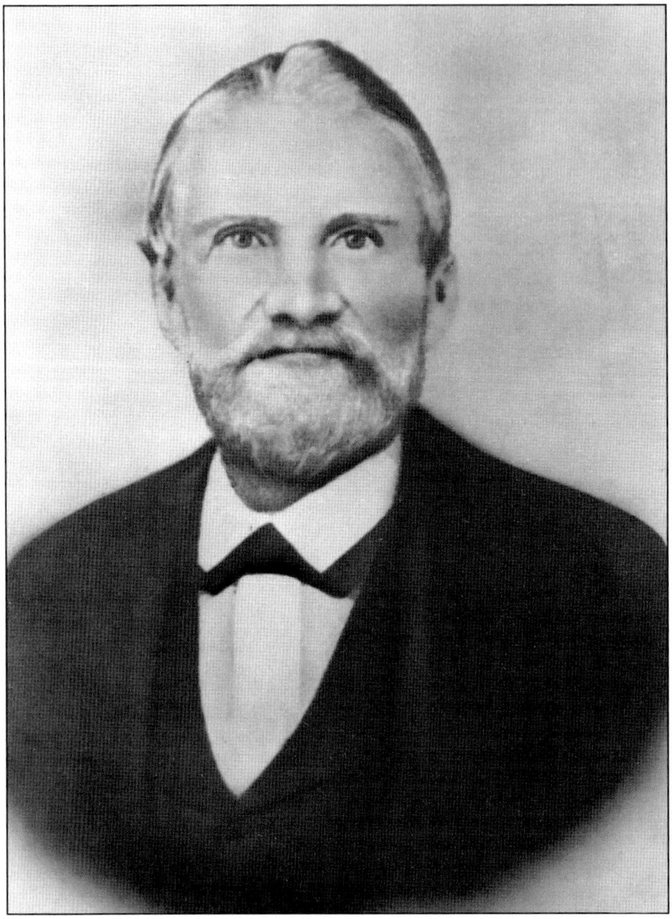

Capt. Fredrick Tresca of France often stopped at Indian trading posts along the Manatee River on his nautical route from Tampa to Key West. His bartering familiarized him with native sign language. Tresca was a great voyager; at 12, he served as cabin boy on the ship that took Napoleon Bonaparte to exile on Saint Helena Island. Many of Manatee's first settlers arrived aboard Tresca's ship, the *Margaret Ann*. (SLAF.)

Here, a dockmaster prepares for the arrival of the steamer *Favorite* of the Favorite Line. H. Walter Fuller of Braidentown, K.W. Wiggins of Manatee, and George Gandy Sr. of St. Petersburg formed the steamship company in 1907. Capt. Ollie Stuart made daily trips to Tampa, St. Petersburg, and Manatee River points, including Palmetto. The trip to Tampa took four hours and included supper. (PHC).

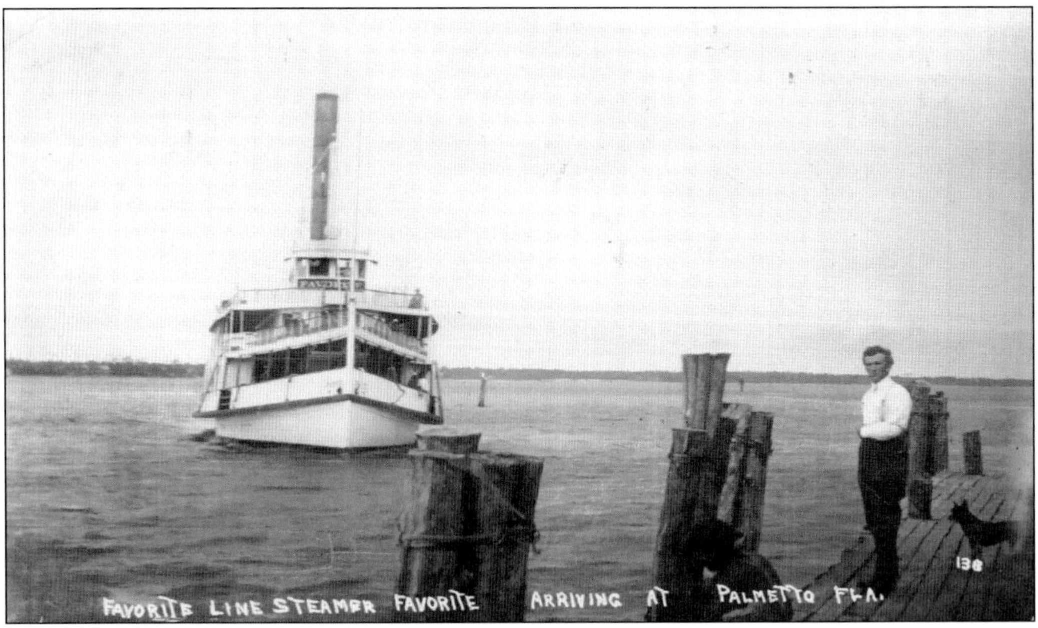

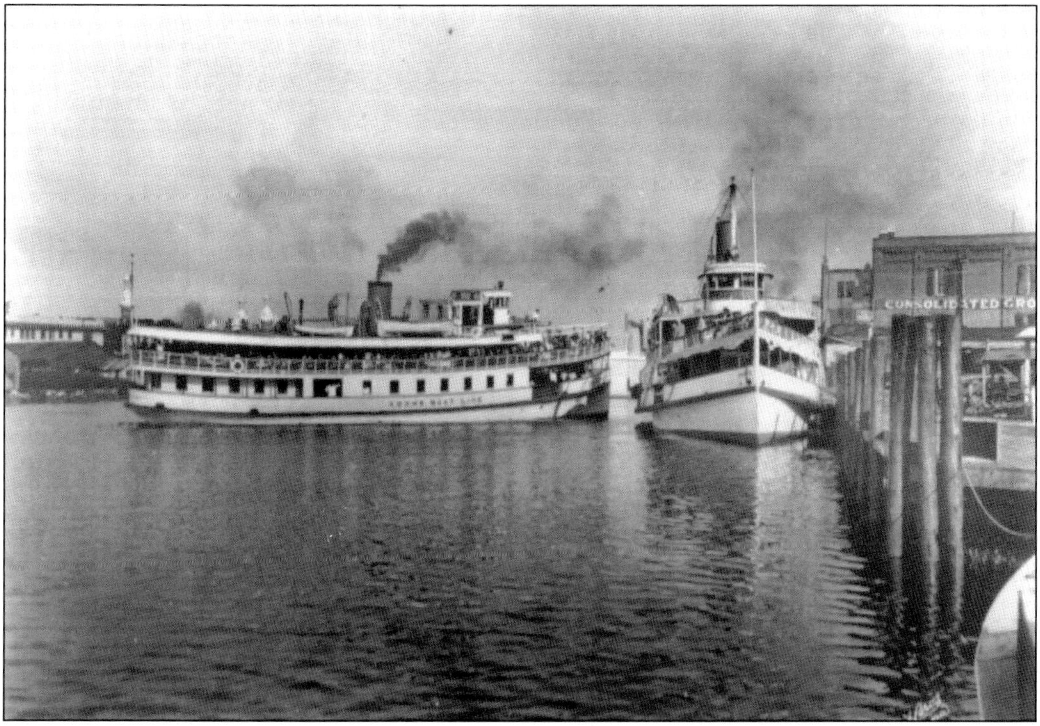

During the summertime, steamers ran a special on Thursdays. For 50¢ apiece, passengers could voyage from Palmetto to Anna Maria in the morning, and on the return route from Tampa another ship would pick them up for the trip home. Girls would pack a lunch for the outing. Steamers *Jeanette* of the Adam's Line and *Favorite* of the Favorite Line approach Palmetto city dock to pick up passengers. (PHC.)

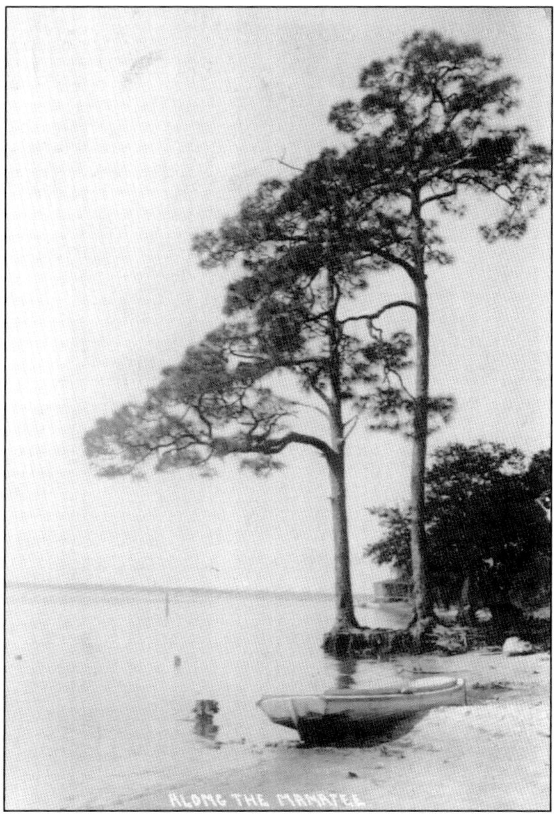

Before bridges were built, the Manatee River separated the towns on the north and south side of the river. Most crossings were done by rowboat. Because both Braidentown (Bradenton's earlier spelling) and Palmetto were so sparsely populated, Palmetto boys would often sabotage the vessels of the young men of Braidentown who left their boats unattended while coming to court one of the few available young women. (PHC.)

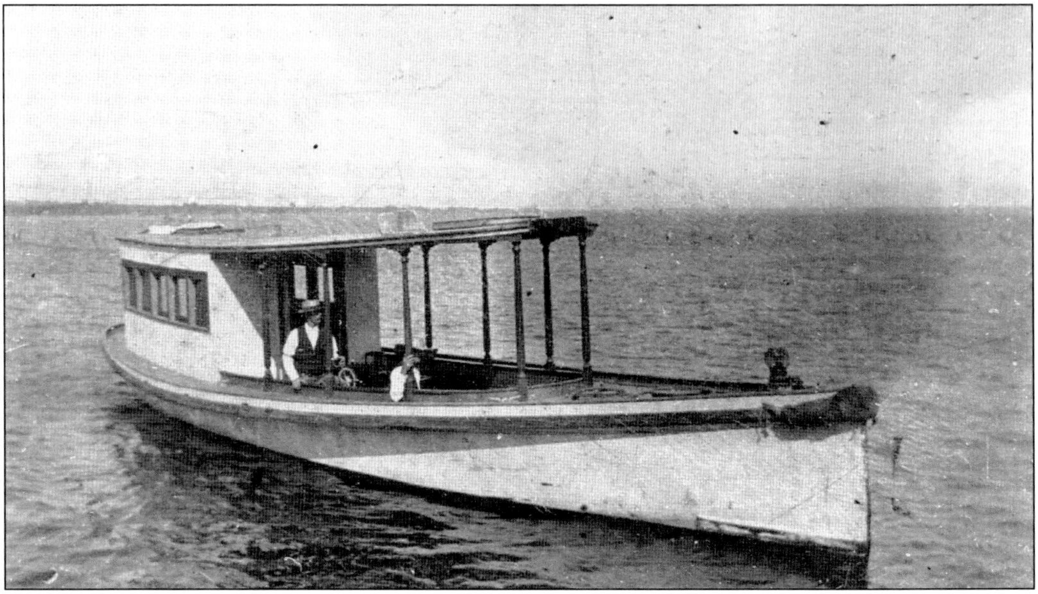

In 1898, Ernest Kretschmar built a gasoline-run launch called the *Zephyr* that could carry up to 25 passengers. The *Zephyr* had a partially enclosed cabin to protect passengers from bad weather and unruly seas. Capt. George Breeze departed from Palmetto and Braidentown every half hour for a fare of 10¢ a person. The launch ran until 1919, when Victory Bridge was built. (MCHS.)

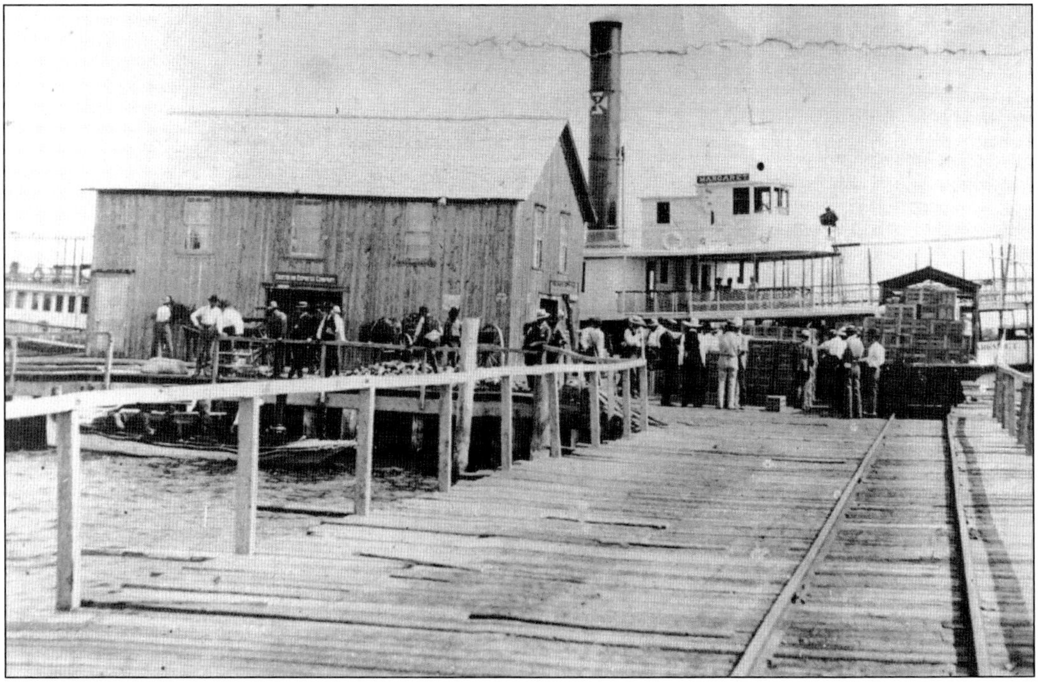

Men on the Hendrix Dock wait to load freight aboard the steamship *Margaret*. Joel Hendrix built the first wharf in Palmetto. The Palmetto Terminal Railroad ran a span of 12 miles from the dock to Parrish and hauled produce from the country farms to dockside, where it was loaded aboard ships. The wharf was not sturdy enough to hold a freight engine, so carts were pushed to the end. (MCHS.)

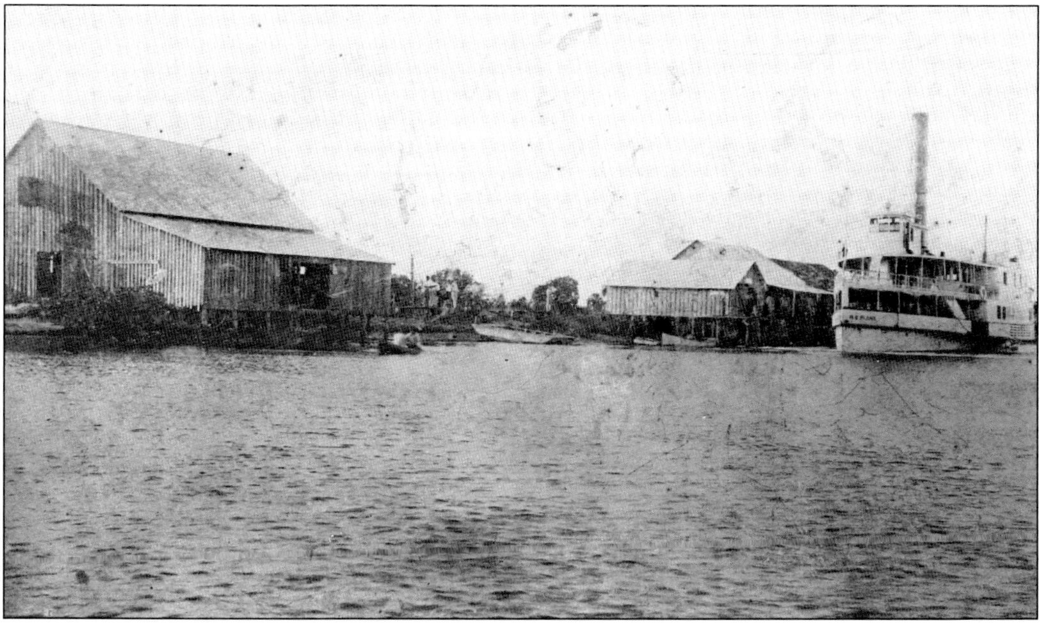

The *H.B. Plant,* named after railroad tycoon Henry Plant, is at the loading dock on the Snead Island Cutoff. The Plant Investment Line owned the boat until 1903, and the Favorite Line purchased it in 1912. It burned at a Tampa pier in 1913. In 1902, Plant had plans to buy Snead Island in its entirety and develop it into a tourist resort, but landowners refused to sell. (SLAF.)

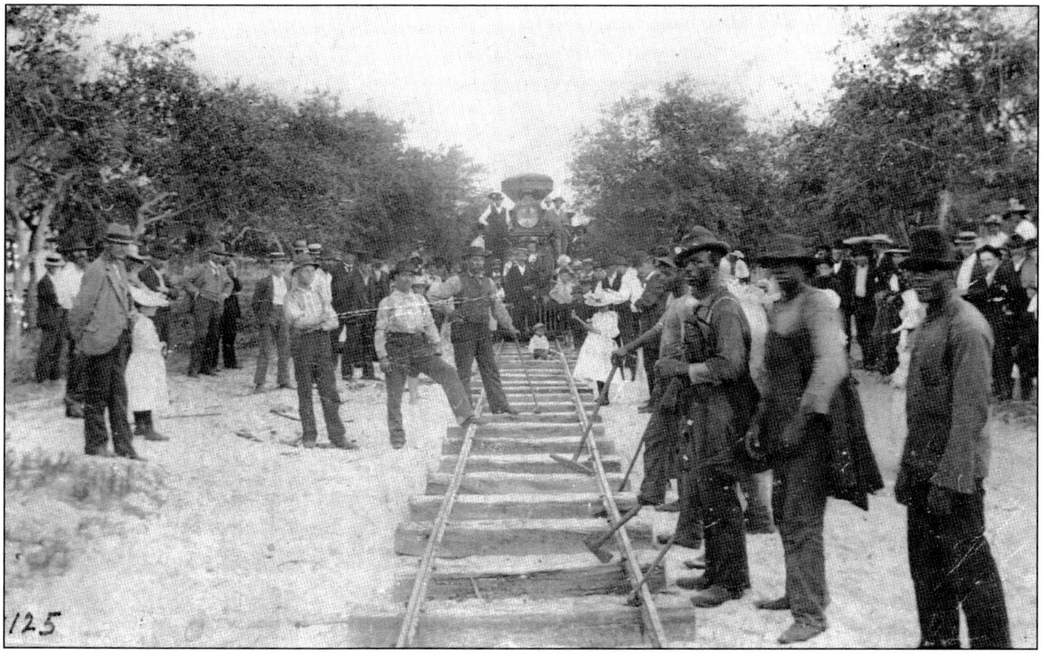

The dedication ceremony for the Palmetto Terminal Railroad was held on April 27, 1895. Annie Lamb (on the right side of the tracks, wearing white) drove in the last gold and silver spike. The $10,000 venture was ballyhooed as a boon to the community, but it had a short life. The project was never completed after the engine died, and carts were pulled with horsepower. The rail spur was eventually abandoned. (PHC.)

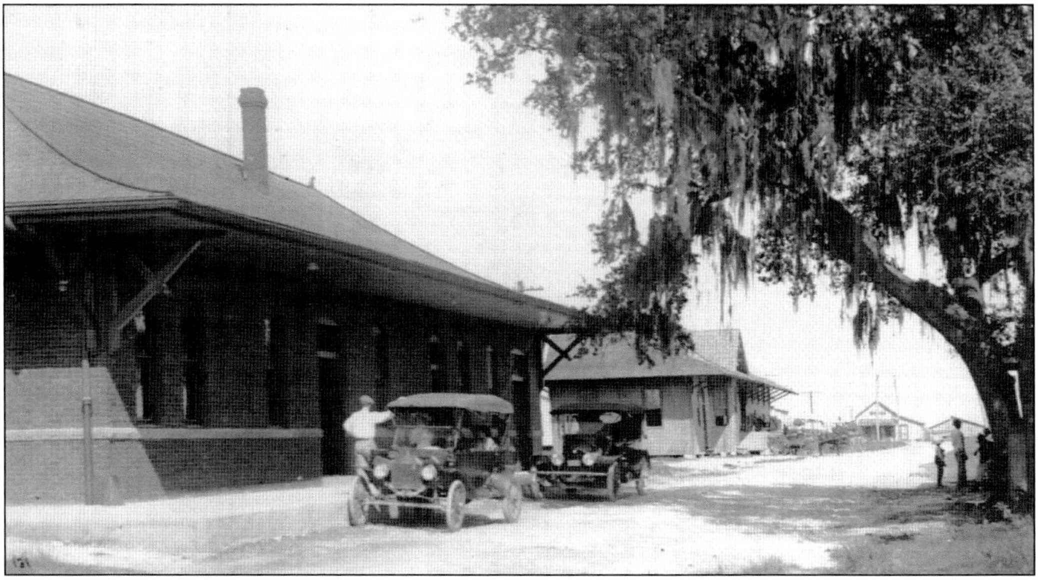

The first railroad that ran from Tampa to Palmetto was built by the United States West Indies and Steamship Company and was purchased by the Seaboard Air Line Railroad (SAL). The first train came into town on May 20, 1902. A spur of the railroad extended to Terra Ceia and Snead Island to serve growers. Pictured here is the Tampa Southern Railroad depot, located on Fourth Street and Fifth Avenue in Palmetto. (PHC.)

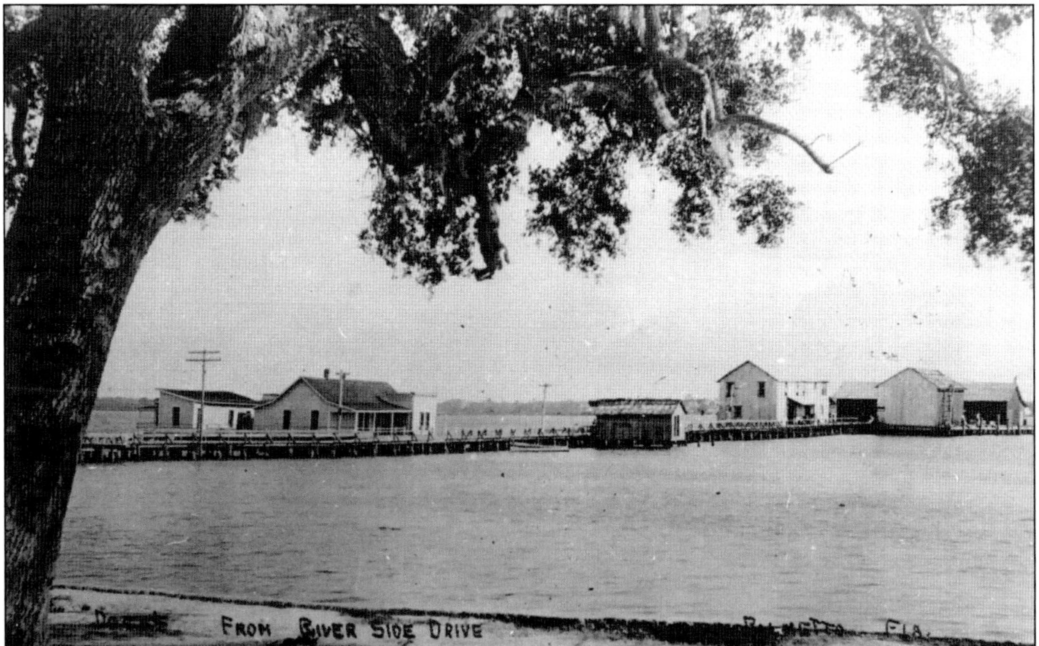

The Palmetto Wharf Company, owned by S.S. Lamb, J.A. Howze, and J.A. Lamb, built City Dock in 1897. The dock was at the end of Tenth Avenue (Main Street). A hurricane in 1921 demolished most of the pier. It was restored, but shipping traffic ceased when railroads and roadways became the favored means to carry goods to market. The pier ended up as a place to fish and moor boats. (PHC.)

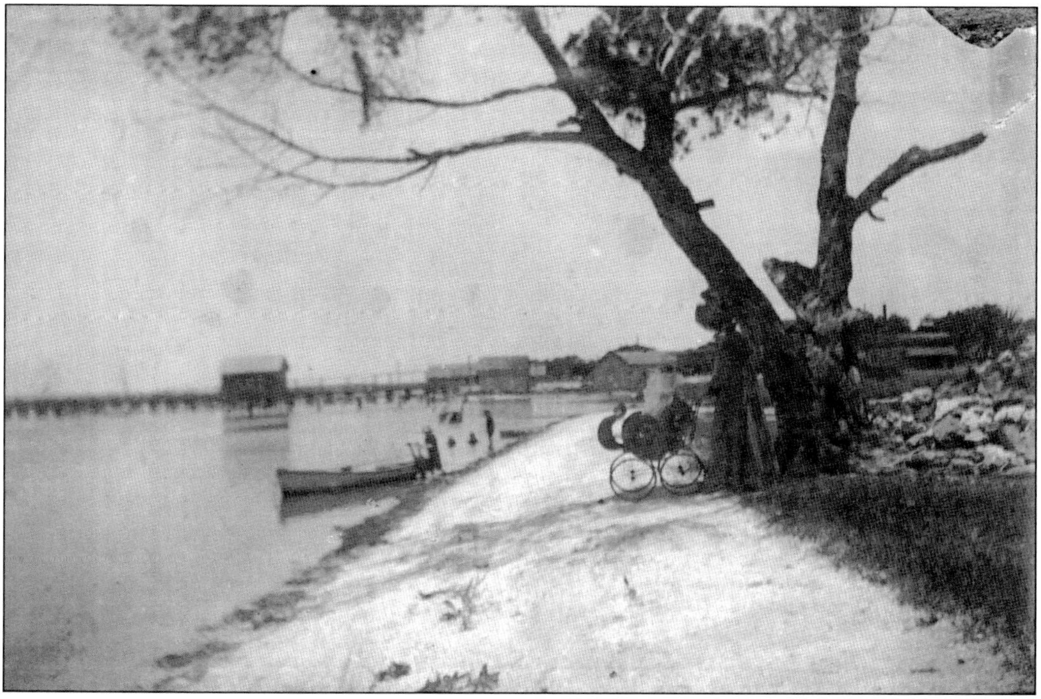

Mrs. T.A. Howze poses with son Paul in the baby carriage. In the background, Palmetto's first jail (known as "Greco's Hotel") can be seen along with three docks: the ice dock, the SAL dock at the foot of Thirteenth Avenue, and the Howze Dock. The ice plant is partially in view to the right of the cedar tree. (PHC.)

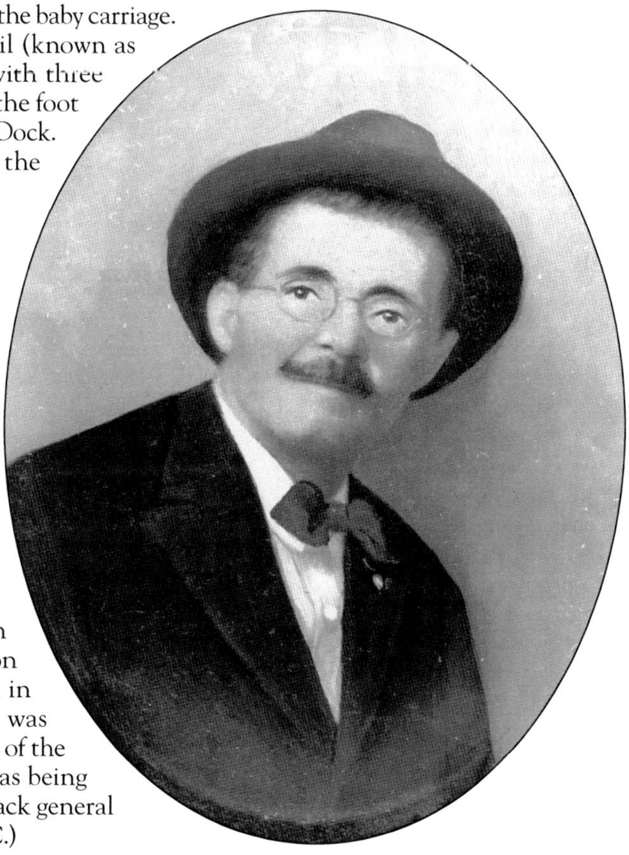

Henry Clifford Webb was an inventor who revolutionized the way land was cleared. After moving to Palmetto in 1907, he patented the Webb Plow on May 15, 1917; it became instrumental in the construction of US 41. The plow was designed to cut through the thick roots of the saw palmetto plant. Webb is credited as being the first black inventor and the first black general contractor in Manatee County. (PHC.)

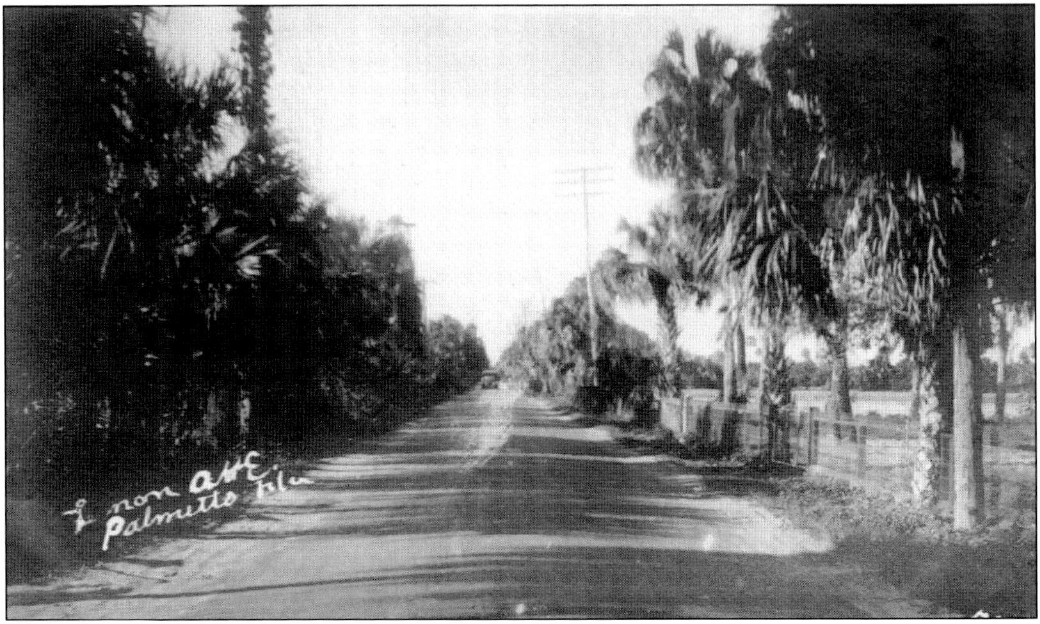

The first official city street to open in Palmetto was Frog Street (now Eighth Avenue). It was changed to Lemon Avenue when the Manatee Lemon Company was established north of town and Frog was the only access. Pictured here, Lemon Avenue was not utilized extensively until the opening of Victory Bridge in 1919. (PHC.)

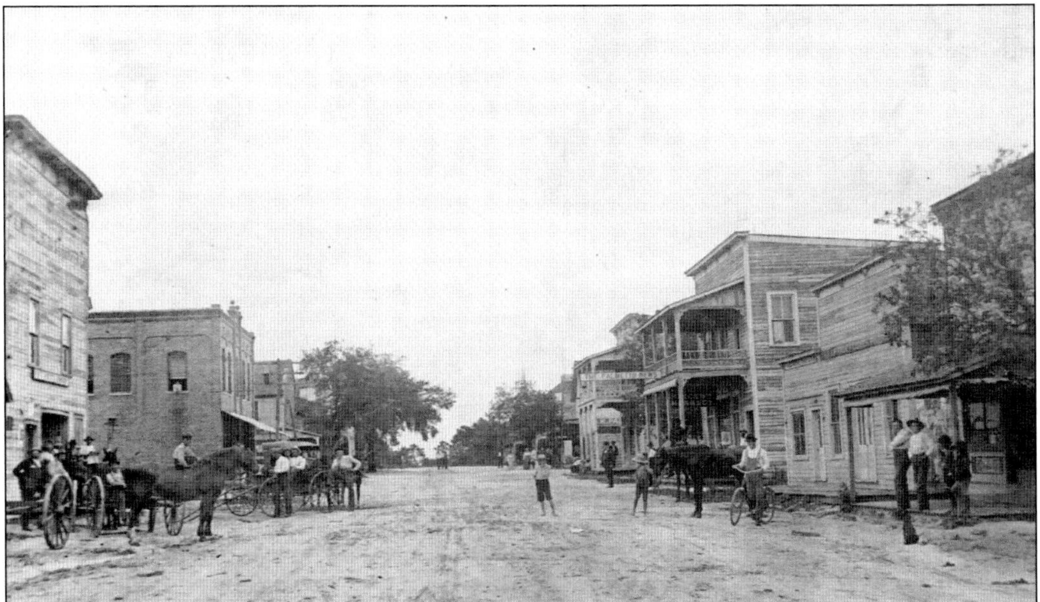

Samuel Sparks Lamb purposely designed Main Street to be extra wide, as he imagined a grand thoroughfare to the river. The upstairs of the second building on the right was the headquarters of *Palmetto News*, founded in 1894 by J. Ira Gore. The newspaper provided news to the community for more than 72 years. (PHC.)

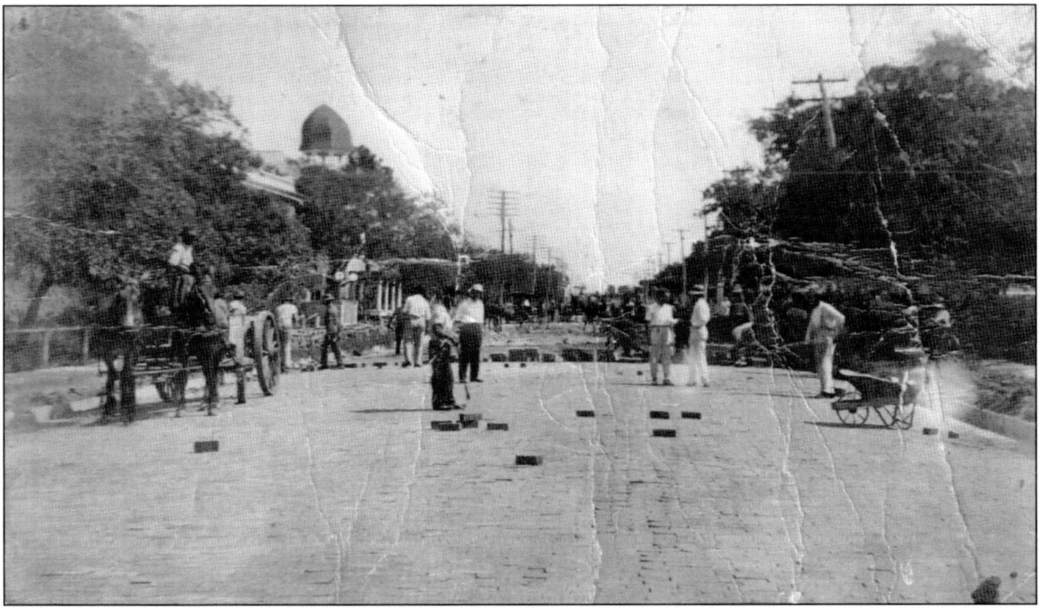

Workers lay out bricks to pave Main Street and to upgrade the shell streets. In 1909, a city surveyor was hired to plat streets, and the city began to undertake civic projects. In 1904, the Village Improvement Association funded $100 for a sidewalk that ran the entire length of Main Street from the river to the SAL depot. (PHC.)

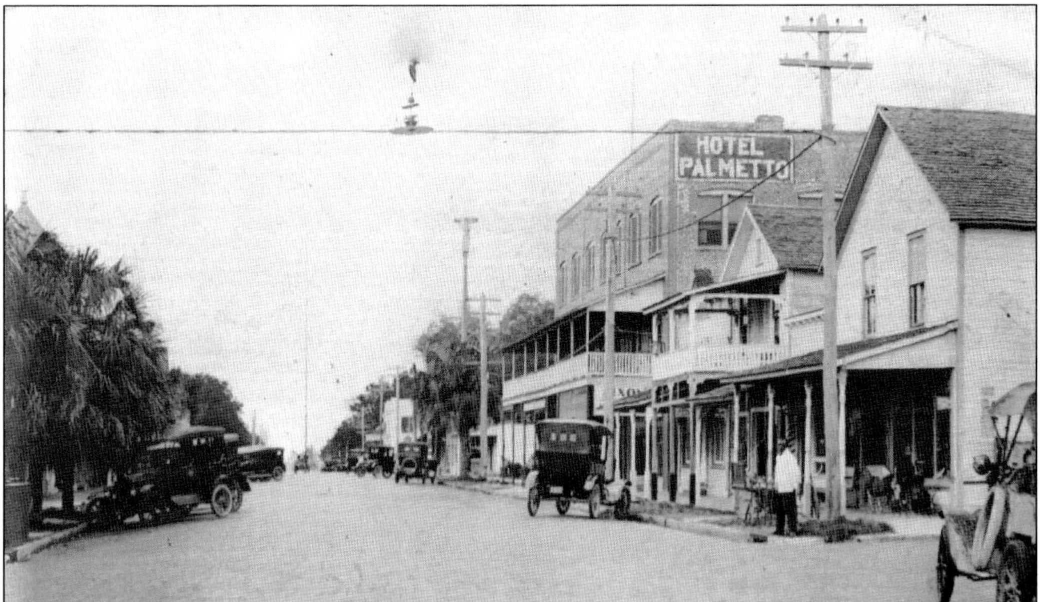

After the paving of Main Street was completed, the city passed an ordinance to keep roaming horses off the newly surfaced streets and sidewalks, and it built its first parking lot—a lot for horses. It was for cars only on Main Street. (PHC.)

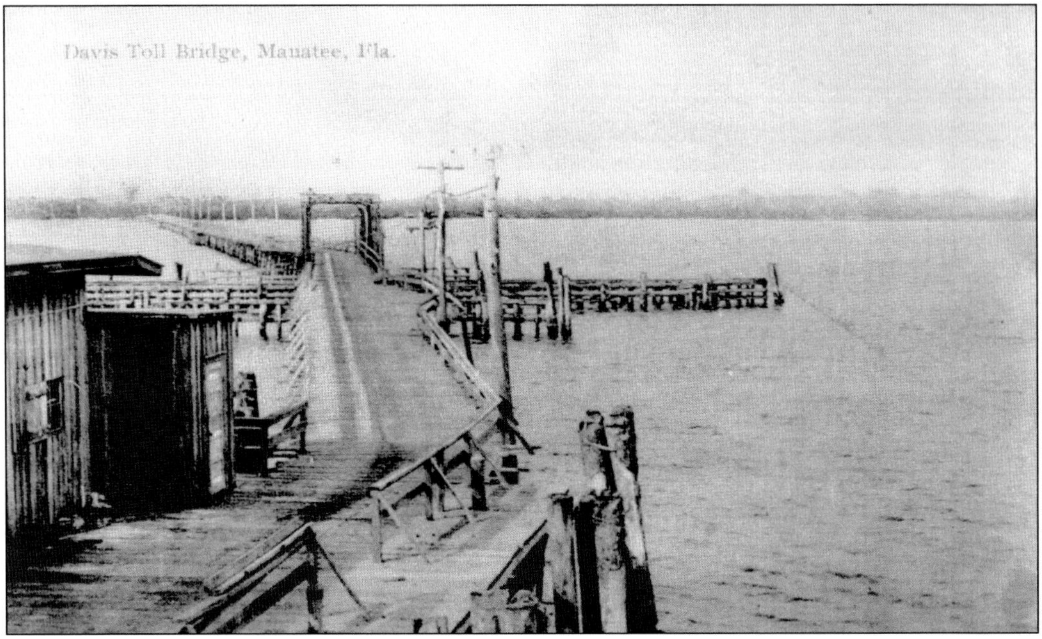

In 1910, the Davis Toll Bridge was privately built by C.H. Davis and opened for public use. It was the first bridge to span the Manatee River. It ran from east Palmetto to the village of Manatee. Cattle traveling over the bridge were blindfolded to prevent them from being spooked. Turnouts were constructed so two wagons could pass each other. (PHC.)

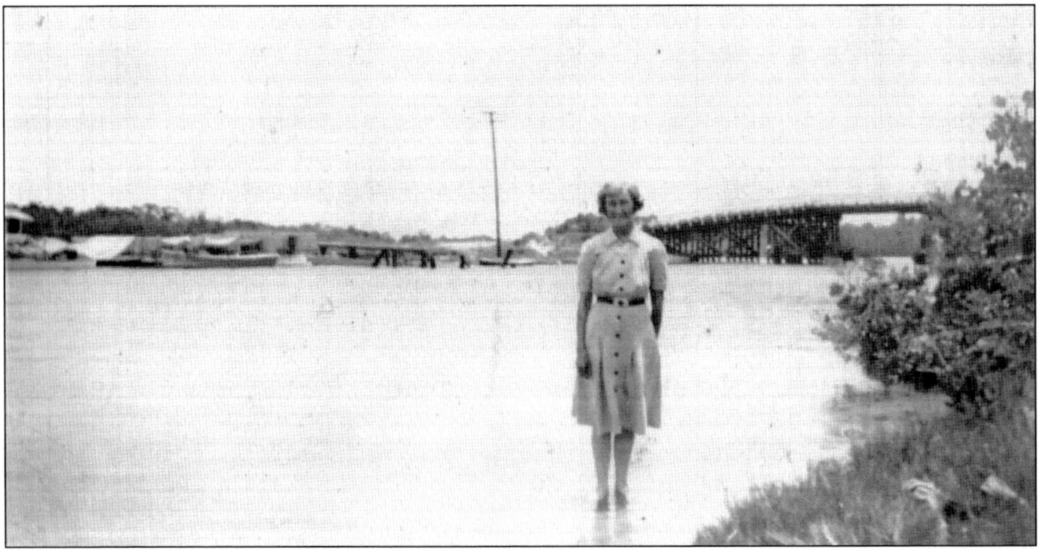

Prior to 1898, a shallow channel separated Snead Island from the mainland—residents usually waded across. In 1898, the government dredged the passage, making steamer travel more efficient. At first, a ferry transported people and wagons across. In 1920, the Davis Toll Bridge was disassembled and reused as a swing bridge for the Snead Island Cutoff. A woman stands in front of the wooden "Humpback Bridge" that replaced it. (PHC.)

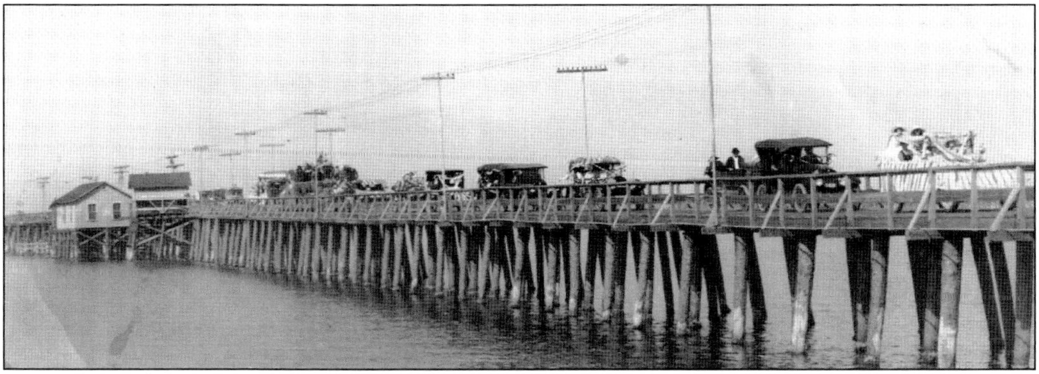

The Palmetto City Commission authorized construction of the first bridge linking Palmetto and Braidentown on May 17, 1917. It ran from Eighth Avenue in Palmetto to Ninth Street in Braidentown and had a drawbridge in the center to allow passage by traveling steamers. Because its construction was completed in October 1918, shortly after the end of World War I, it was named Victory Bridge. At its opening, automobiles paraded across the span in celebration. The bridge was barely wide enough for two cars to pass, but residents did not mind—they finally had a free bridge. (Courtesy Nick Baden.)

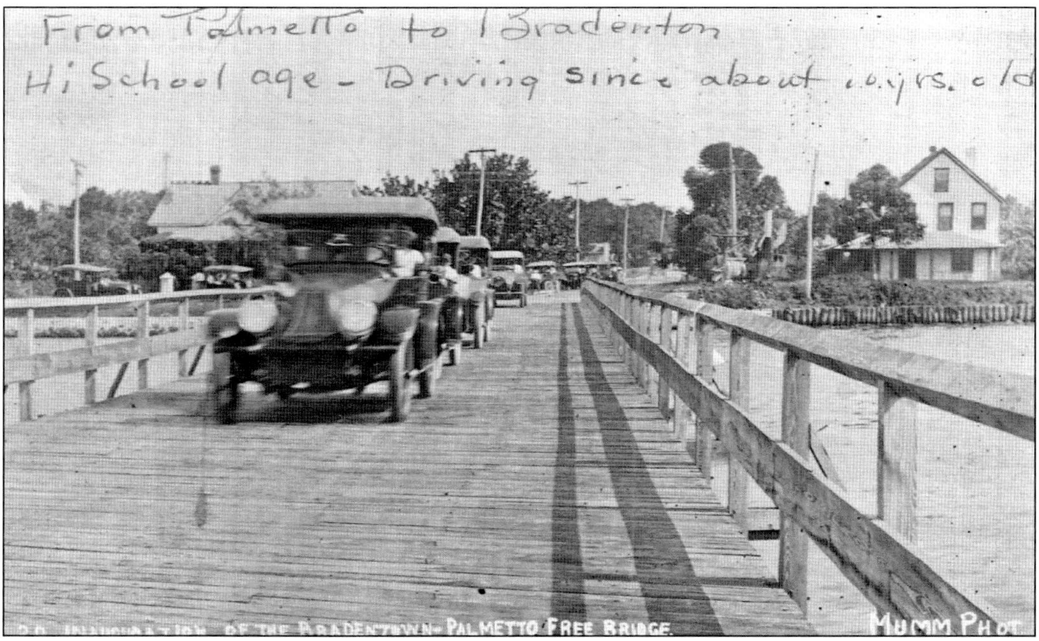

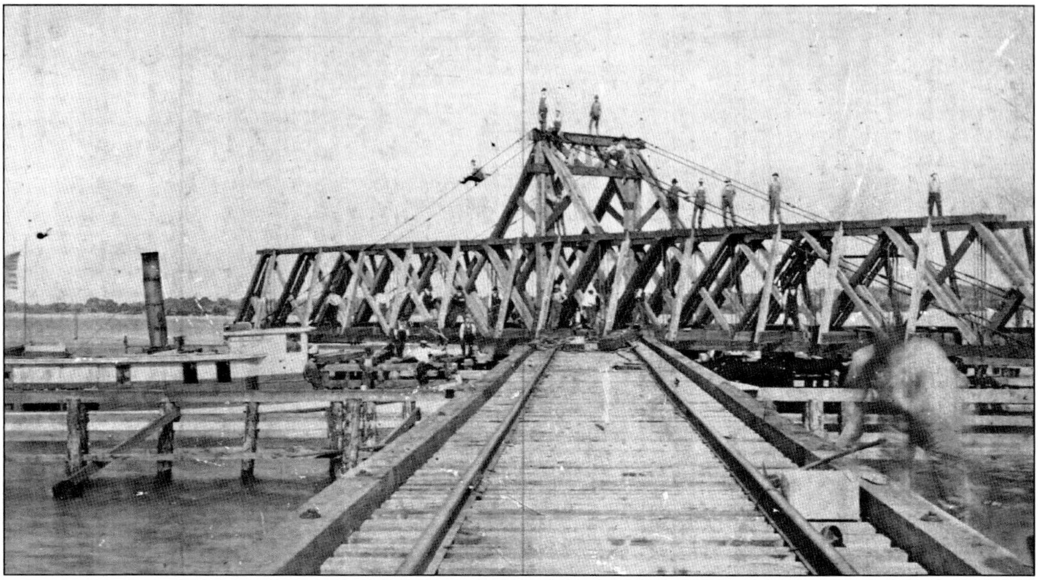

In 1919, the Tampa Southern Railroad finished construction on a span connecting Palmetto to Braidentown. It was located just the east of Victory Bridge. Workers are seen on top of the swinging trestle that allowed for ship passage. Tampa Southern Railroad became the Atlantic Coastline Railroad before finally becoming the Seaboard Coastline Railroad. (PHC.)

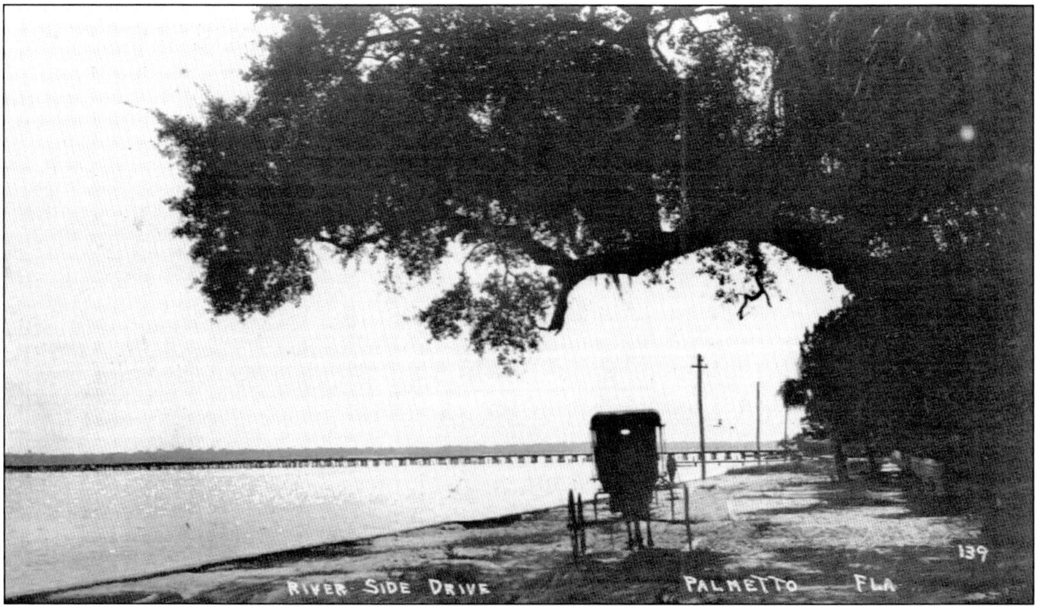

A horse-drawn buggy travels down an oak-shaded Riverside Drive; Victory Bridge is visible in the background. When this photograph was taken in 1918, the road was nothing but a dirt path. (PHC.)

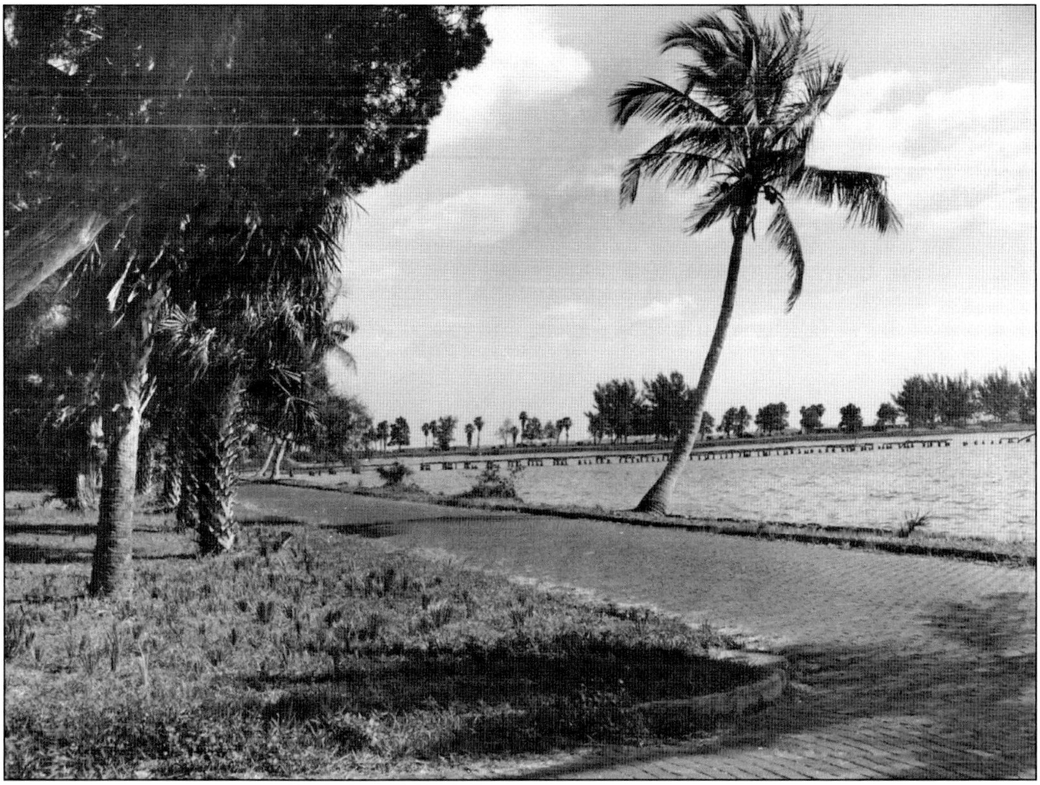

After automobiles became the preferred means of transportation, Riverside Drive was paved with brick. The street was more residential than any of the other main thoroughfares, as is evident in this landscape photograph by Harold R. Smith. (PHC.)

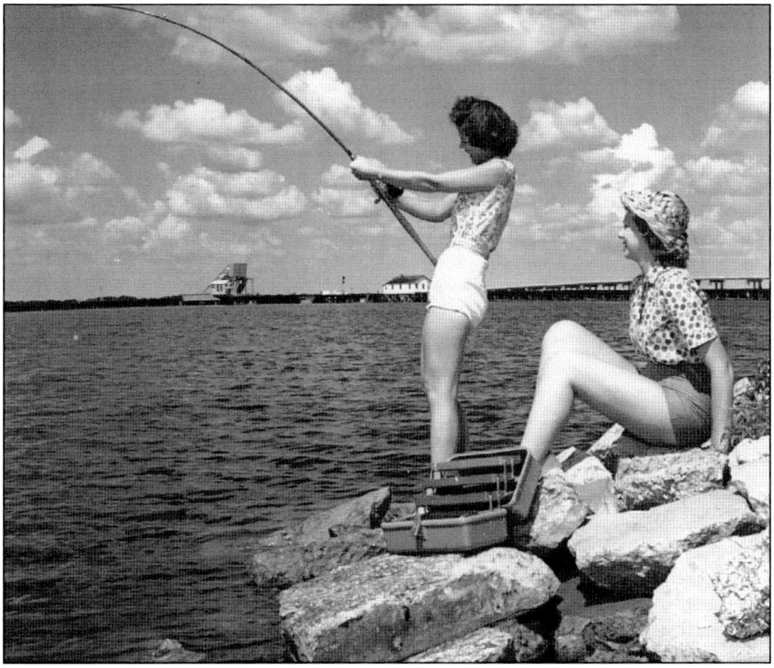

Women fish off the riverbank with Green Bridge visible in the background. Due to unbearable traffic congestion from the narrow Victory Bridge, E.P. Green of Bradenton lobbied for a new state-funded bridge to be built alongside the old structure. The $1 million project began on January 18, 1925, and when the new bridge opened on June 1, 1927, it was named after its longtime advocate. (PHC.)

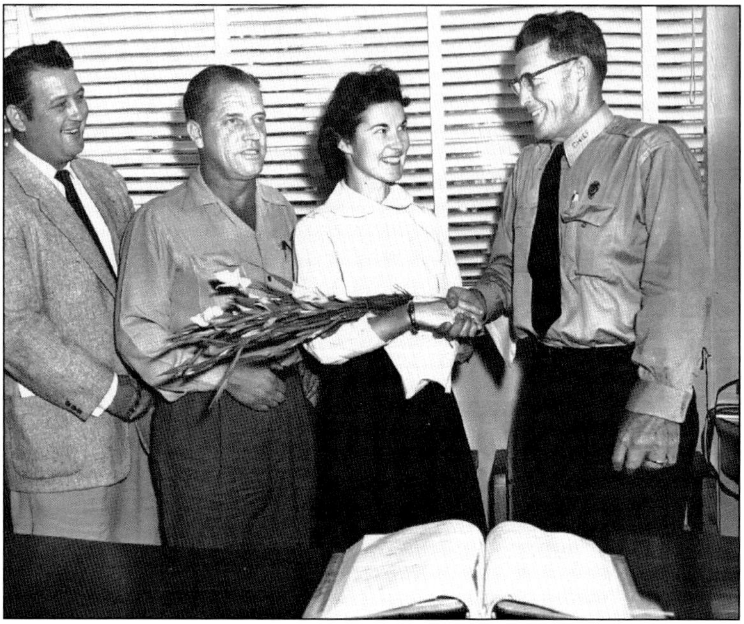

During the 1950s, AAA would route drivers around Palmetto because of the city's reputation as a speed trap. In an effort to put Palmetto in a more positive light, the Palmetto Chamber of Commerce instituted "hospitality traps." At random, travelers were pulled over and sent to "hospitality court," where they received bouquet of gladiolas and gratis hotel and dinner accommodations instead of a ticket. (PHC.)

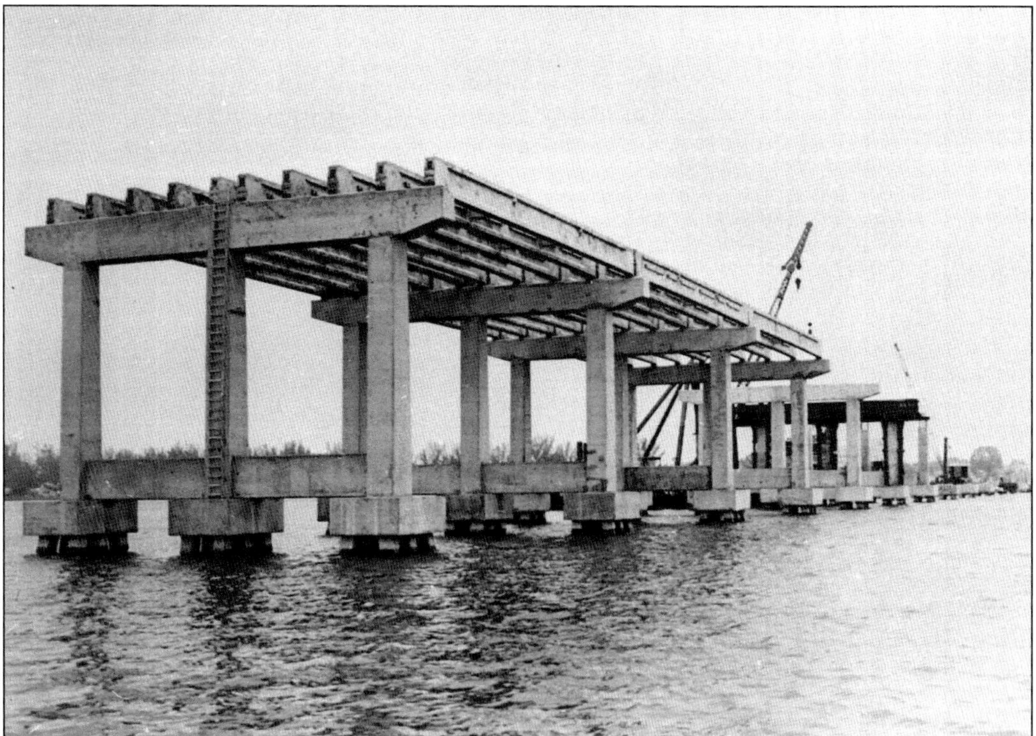

In 1957, the state built DeSoto Bridge to be used as part of US 301. Instead of having a draw in the center, DeSoto Bridge featured a high span to accommodate river traffic. A water view taken during the bridge's construction looks north toward Palmetto's shoreline. (MCHS.)

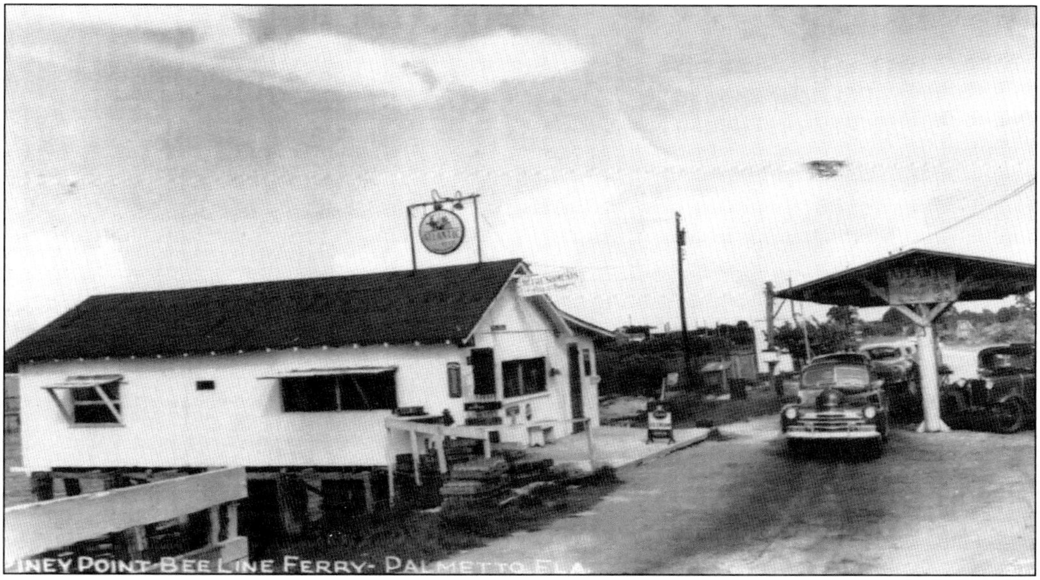

The Bee Line Ferry, with its terminal north of Palmetto at Piney Point, reduced a 69-mile road trip from Manatee to Pinellas County to just 22 miles—a scenic trip that took only an hour. Launched in 1926, the ferry service was hailed as a transportation breakthrough. Three vessels ran trips every 30 minutes from 8:15 a.m. to 8:45 p.m. A $3 special included a round-trip ticket for a vehicle and its passengers. In 1942, the privately owned company ceased operation when the federal government requisitioned the boats for the war effort. Motorists had to detour 50 miles inland until the Sunshine Skyway opened in 1954. In its final year, the ferry carried 99,000 vehicles and grossed $170,000 in revenue. (MCHS.)

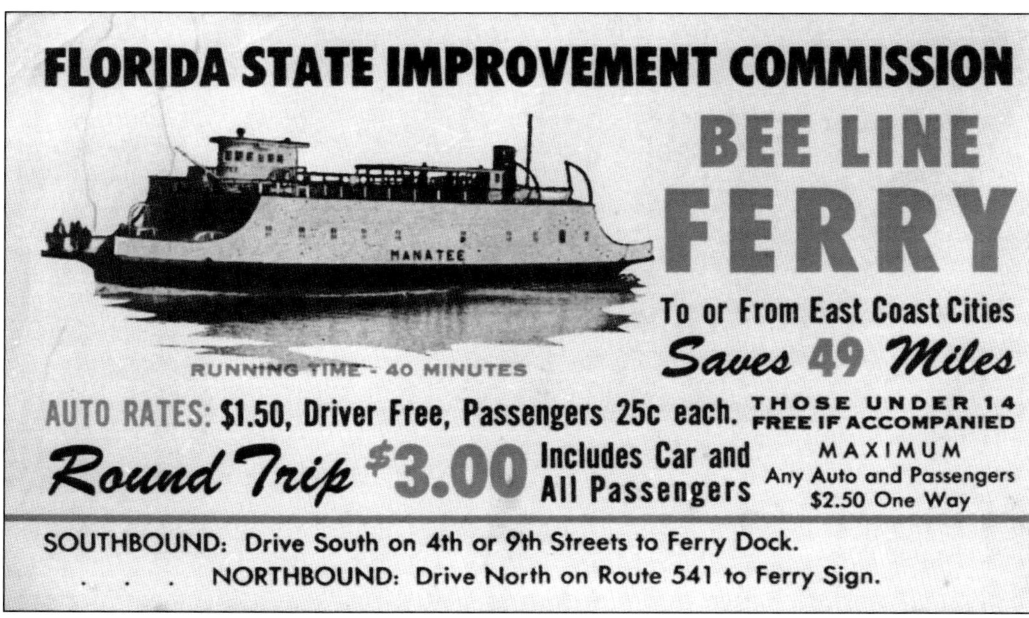

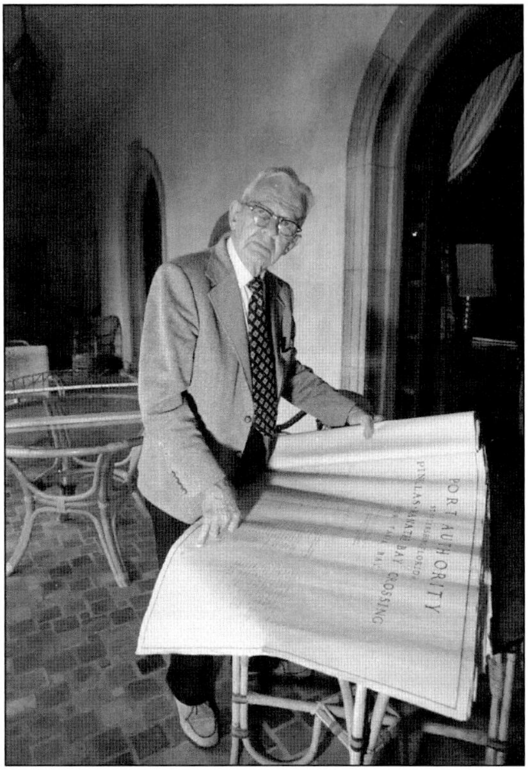

In 1918, Freeman H. Horton (born August 13, 1897, in Fogartyville) became the first Manatee County High School student to graduate in civil engineering from Harvard University. In 1924, he moved to Snead Island and formed an engineering consultant company called Bail, Horton and Associates. On December 20, 1944, the firm received a contract to design the Sunshine Skyway. Under his original plans, Horton proposed Snead Island as the southern terminus, with Tenth Street becoming the main thoroughfare. The plan would have permanently split the city of Palmetto. The state suspended construction for lack of funding. In the early 1950s, officials awarded the final project to a New York engineering firm that shifted the landfall to Terra Ceia. The steel, two-lane cantilever bridge, spanning 5.5 miles across Tampa Bay, was completed in 1954 for $21 million. (MCHS.)

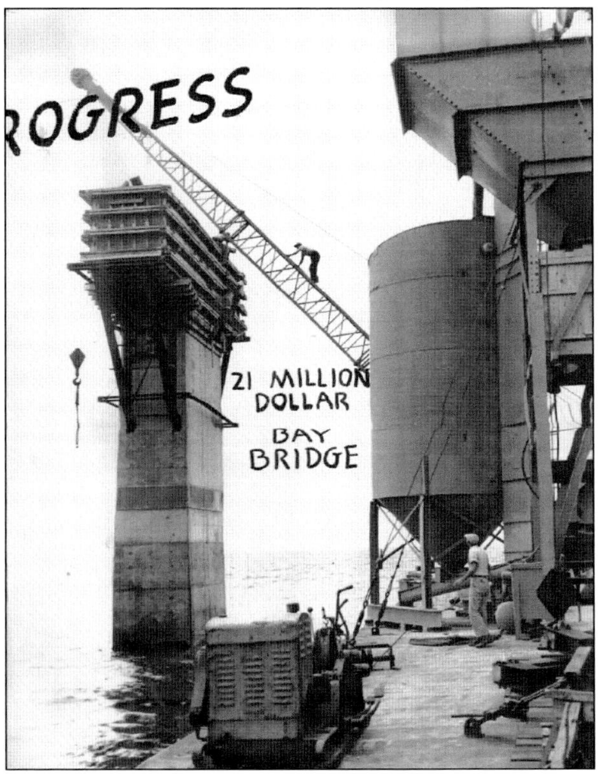

Three

Agriculture
Innovative and Bountiful

Beginning with the early native dwellers, fishing has been considered Palmetto's first industry. The natives made nets woven of palm fiber. Much later, during the Cuban trade period, Spanish-style nets constructed of cotton twine replaced the original material. After the Civil War, a wave of commercial fishermen settled on Florida's southern coast and established an enterprise that continues today. In 1902, a particularly bountiful year, Florida's west coast produced more than 48 million pounds of seafood.

Early pioneers did a little bit of everything: they fished, farmed, and raised and sold cattle. Capt. James McKay, an entrepreneur from Tampa, is credited with starting trade with Cuba in the late 1850s, shipping up to 400 cows a month, until the Union blockade during the Civil War thwarted exports.

The history of organized agriculture in Palmetto begins with sugar cane, which then led to celery crops, followed by truck vegetables, fruit, citrus, and finally tomatoes. After the area opened for settlement in 1842, the mild climate and rich soil attracted people of sufficient means to engage in large-scale cultivation. Today, Palmetto remains one of the top tomato producers in Florida.

At first, produce grown for market had to be shipped to Cedar Key. Joel Hendrix, who settled in Palmetto in the 1870s, is known as the father of green-vegetable shipping. Through experiment, he determined the number of days it would take to ship unripe fruits and vegetables by boat from the Palmetto wharf to Cedar Key, then by rail to New York, arriving ready for consumption. In 1902, the H.B. Plant Railroad extended to Palmetto. The rail connection spurred an immense increase in agricultural growth and profits—enough to necessitate the opening of the first bank in Manatee County. Later, innovative techniques developed by local farmers put Palmetto on the map for a number of agricultural achievements.

Sharecroppers in Manatee County contributed to agricultural production as well. Some growers set aside land for segregated farming communities, such as nearby Rubonia. Over the decades, migrant workers also toiled to harvest the crops for which Palmetto is noted.

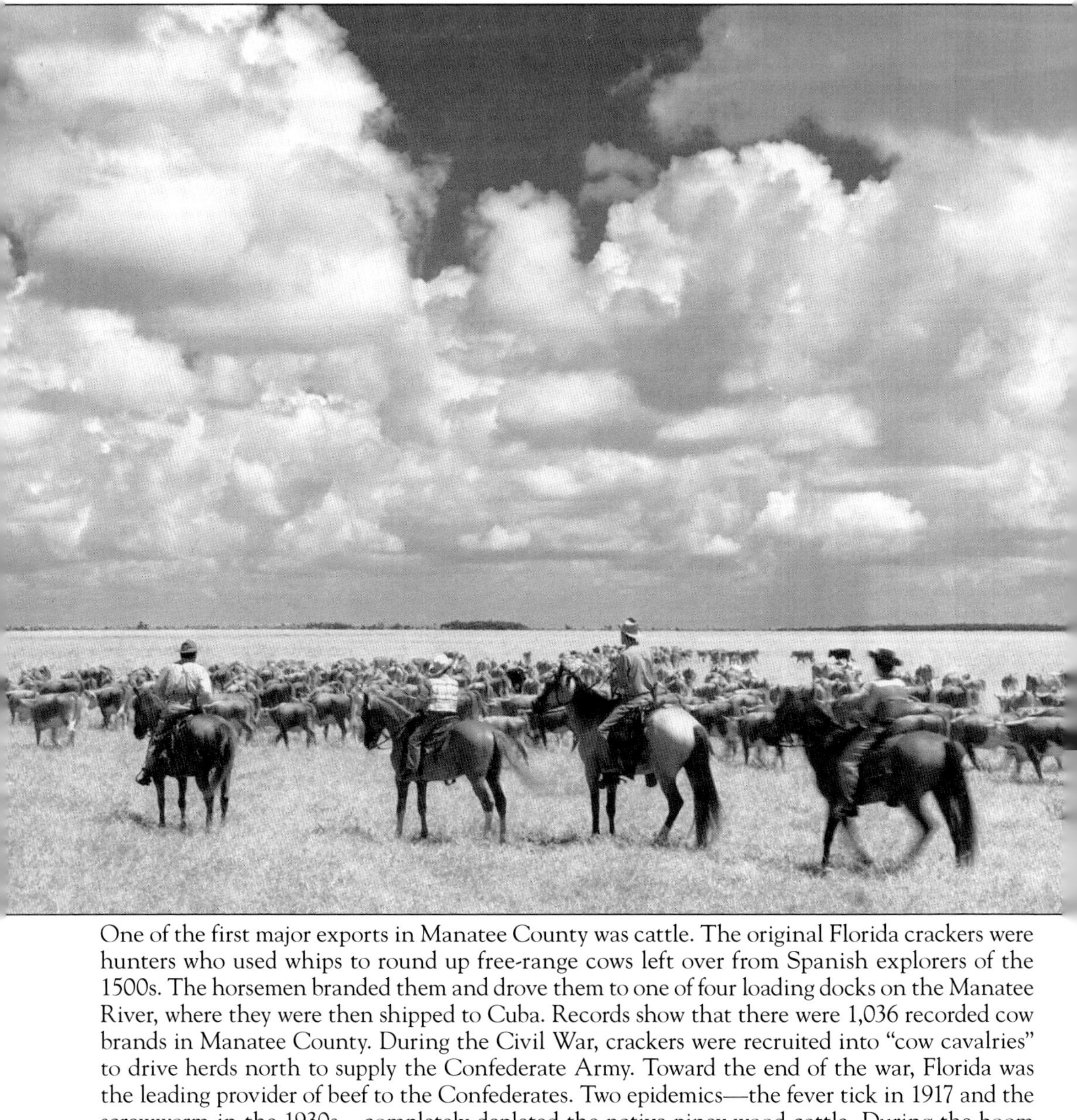

One of the first major exports in Manatee County was cattle. The original Florida crackers were hunters who used whips to round up free-range cows left over from Spanish explorers of the 1500s. The horsemen branded them and drove them to one of four loading docks on the Manatee River, where they were then shipped to Cuba. Records show that there were 1,036 recorded cow brands in Manatee County. During the Civil War, crackers were recruited into "cow cavalries" to drive herds north to supply the Confederate Army. Toward the end of the war, Florida was the leading provider of beef to the Confederates. Two epidemics—the fever tick in 1917 and the screwworm in the 1930s—completely depleted the native piney-wood cattle. During the boom of the 1920s, fences went up all over the state. Although Florida continues to be one of the leading beef producers in the nation, the original crackers became extinct with the demise of the piney-wood cattle and the disappearance of Florida's free range. Here, crackers drive cattle over an expansive prairie. (SLAF.)

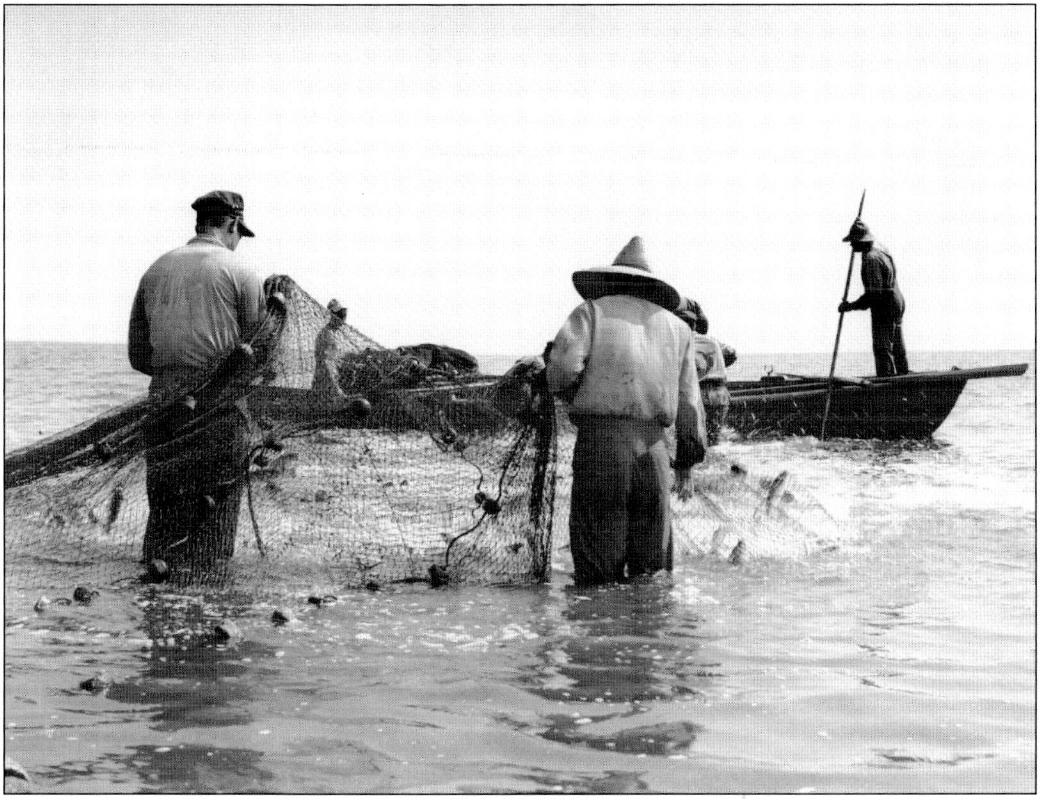

A photograph by Harold R. Smith features several commercial fishermen using gill nets. Nets have been the principal tool for harvesting in local waters over the years. Fishermen traditionally maintained several nets that varied in size, depth, length and cork and lead placement. Each one served a specific use, depending on the time of year and the species of fish sought. (PHC.)

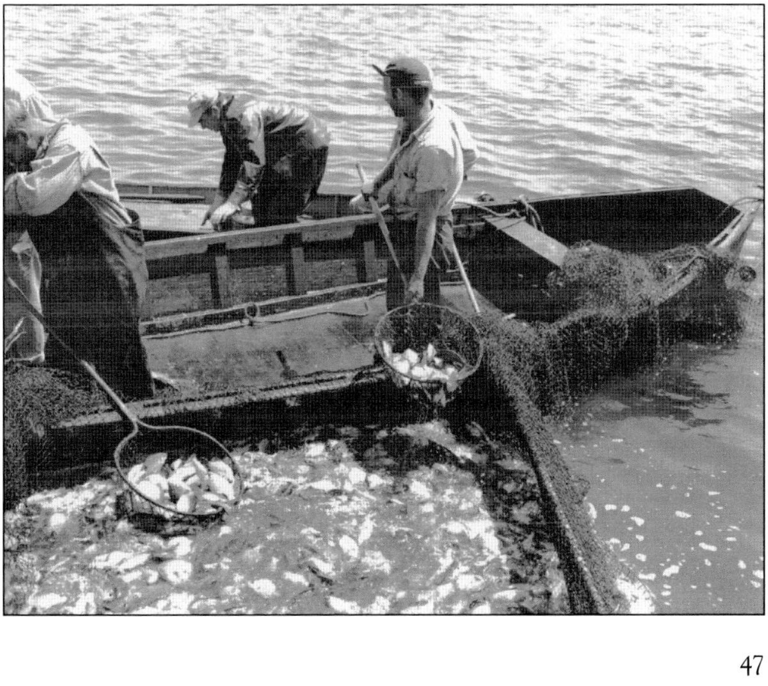

In another Harold R. Smith photograph, commercial fishermen gather baitfish to market. Their boats were typically around 25 feet long and poled out to the flats for fishing, especially when going after an easily spooked species. By the 1930s, gas-powered vessels had replaced the pole boats. (PHC.)

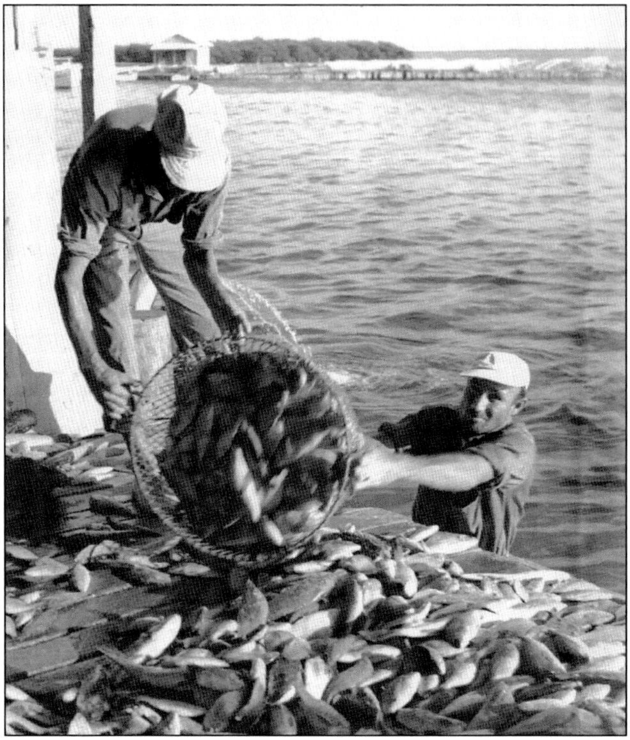

Fishermen unload their catch at a commercial fishing dock in this photograph. Visible in the background are a stilt house and net spreads. Stilt houses provided a place for the fishermen to rest and eat before poling their catch back to the mainland. Lumber had to be floated out to the site, and pilings were set to support them. (PHC.)

Before the invention of nylon, nets were handcrafted of cotton twine. The material, however, would absorb the water and become saturated. Fishermen devised simple structures called net spreads that allowed the twine to dry out under the sun. Local photographer Harold R. Smith likely captured this scene at Terra Ceia, but net spreads could be found throughout the area. (PHC.)

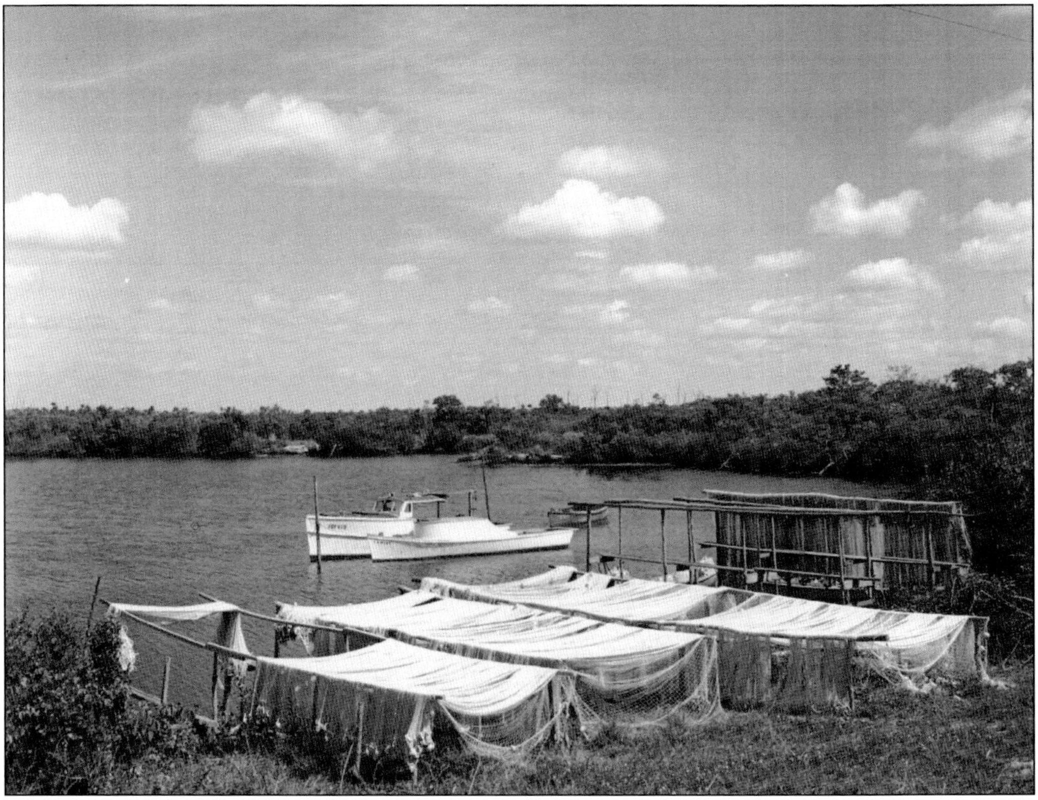

In 1842, Maj. Robert Gamble purchased 3,450 acres of land, planted 1,500 acres of sugar cane, and constructed a mansion built of tabby, located on present-day US 301. By 1850, Gamble's plantation included up to 600 slaves and a processing mill with machinery imported from New Orleans. During the 1840s and 1850s, Gamble was the leading producer of sugar in Florida. Falling prices and mounting debt forced Gamble to sell in 1858. (MCHS).

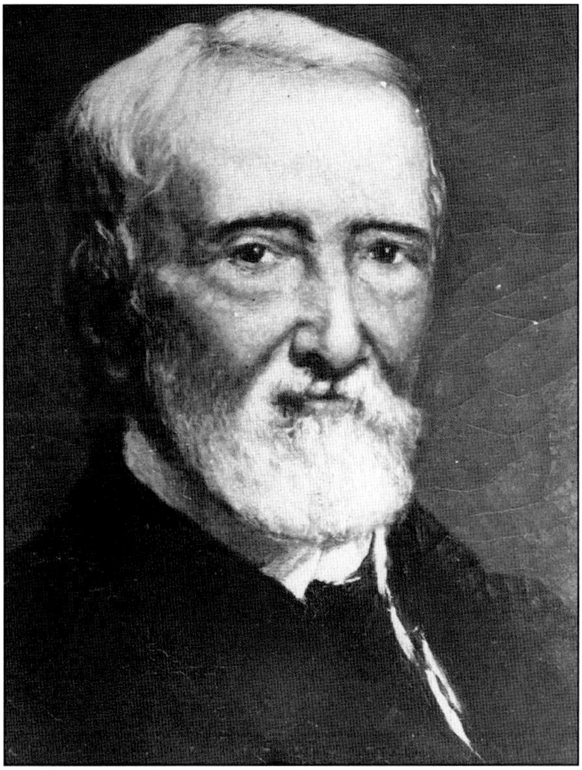

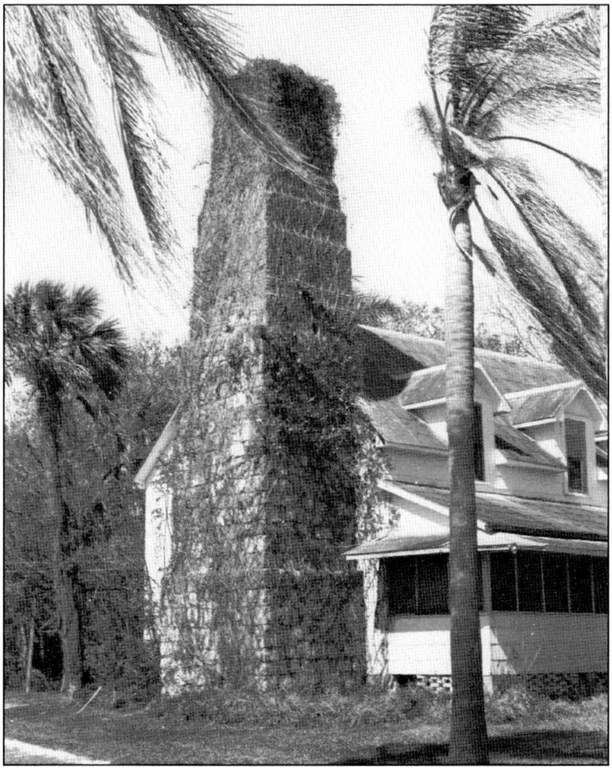

William Pinkston Craig came to the area in the early 1840s to recoup from financial losses after the 1837 collapse of the Union Bank in Tallahassee. He acquired 160 acres. With slave labor, he constructed a sugar mill that had a massive chimney with built-in slots and portholes to allow for defensive gunfire in case of attack. The mill was destroyed during the Civil War. All that remains is the 45-foot-tall chimney. (MCHS.)

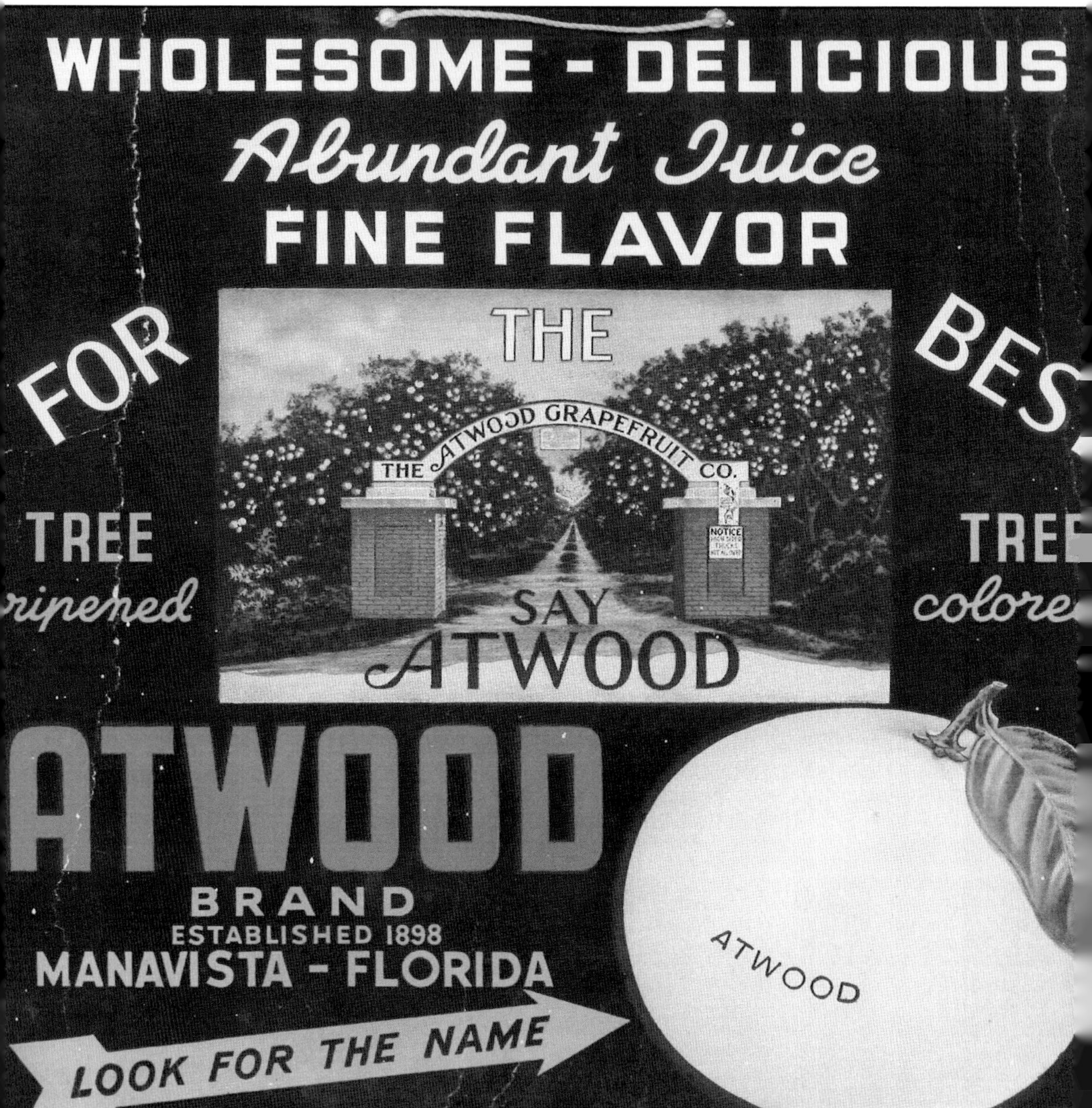

At one time, the largest grapefruit grove in the world was in Manavista, just east of Palmetto. In 1892, Kimball C. Atwood purchased 2,645 acres on the Manatee River and employed 500 men to clear the land. The finished grove was touted to comprise 96 miles of grapefruit. Five artesian wells supplied the grove with irrigation. This sign advertised the Atwood brand, which became popular all over the United States and in Europe. When England's King George V tasted a sample brought to the royal household for a party, he wrote the Atwoods about placing an order. It became an annual tradition to ship the king a complimentary box of handpicked grapefruit. Atwood used to joke it was the least he could do since his grandfather was a captain of a Maine regiment at the Battle of Yorktown. The grove reached its peak in 1927 and remained in Atwood's possession until his death in 1936. His family kept the property until 1968. (PHC.)

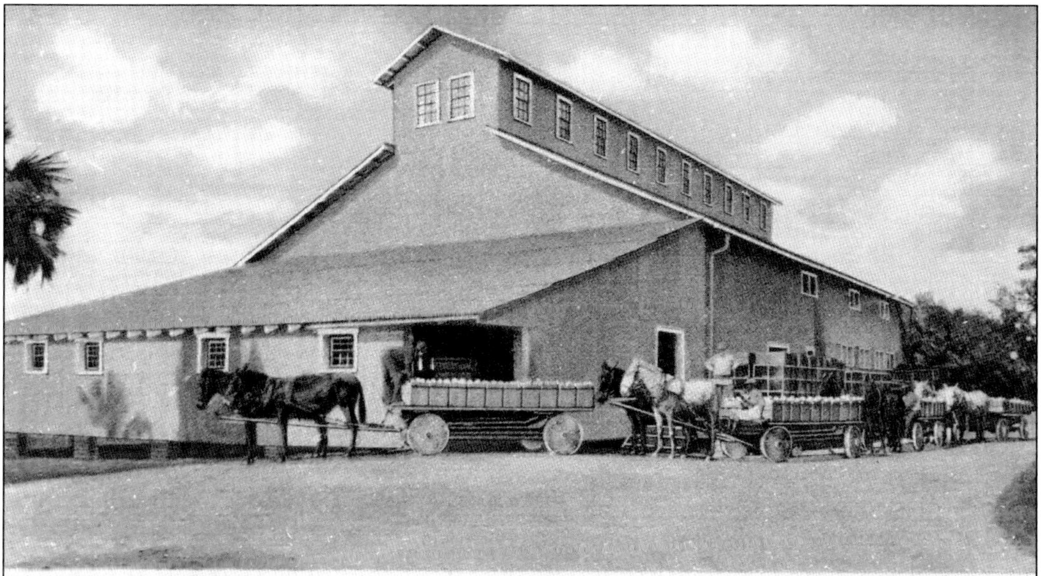

ATWOOD GRAPEFRUIT CO. PACKING HOUSE AT MANAVISTA, FLORIDA — ON THE MANATEE RIVER

At the Atwood Grapefruit Company packinghouse, steamers would pick up cargo at the end of a quarter-mile-long pier, which was wide enough for two wagons to pass. On top of the packinghouse, the cupola served as an observation tower. A lookout would sound a conch shell to alert workers to prepare for an incoming ship. (MCHS.)

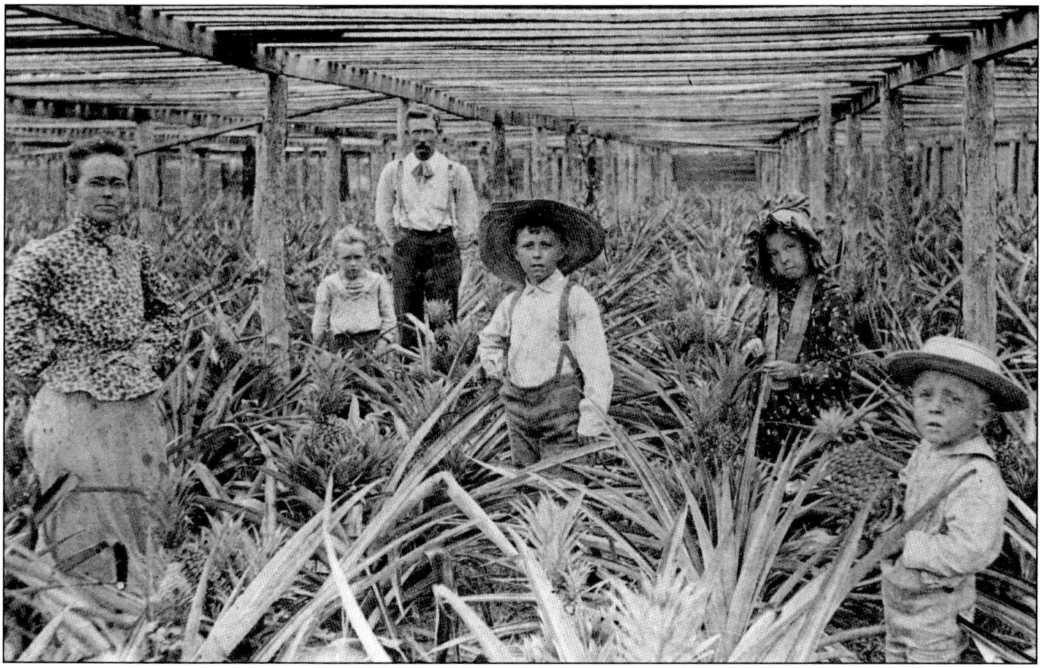

The Pollard family poses at their pineapple farm, known as "The Pinery," on Fourth Street and Seventh Avenue. Talbot Sharp Pollard came to Palmetto in 1895 to recover from pneumonia. When his health improved, he planted crops. The farm was not a success. In addition to competition from growers in the Caribbean and Cuba, his plants proved popular among the area's free-range cattle, which munched on the immature fruit. (MCHS.)

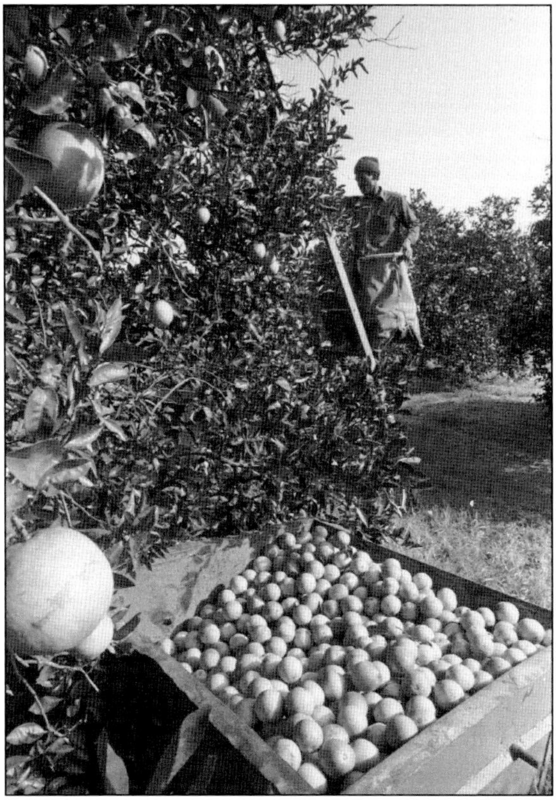

In 1892, a group of wholesalers, led by Julius H. Preston from Providence, Rhode Island, purchased a large tract of land on the Manatee River that extended all the way to Terra Ceia Bay and established the Manatee Lemon Company. Three hundred acres of lemon trees were planted, and, at the time, the grove was considered the largest in the world. With all the picked fruit traveling up Eighth Avenue, the road was renamed Lemon Avenue. The winter of 1907–1908 damaged so many lemons that the company turned to more hardy orange and grapefruit trees. In 1911, the owners changed the firm's name to Manatee Fruit Company, and gladiolas where added as a crop (below). (MCHS.)

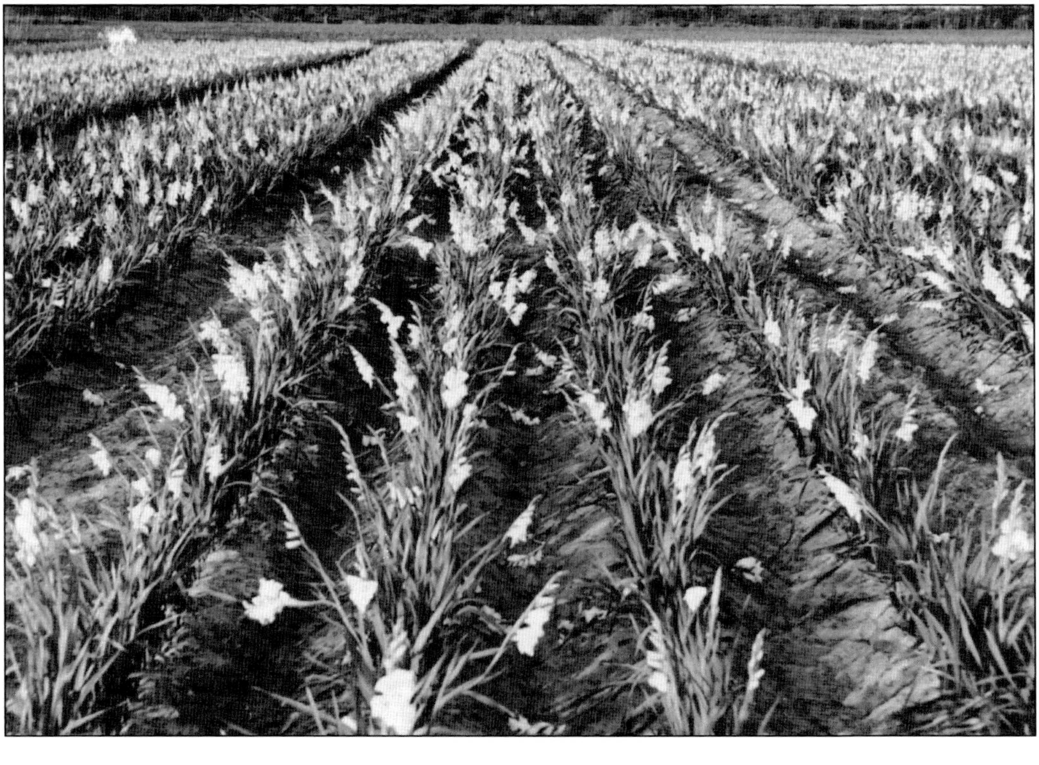

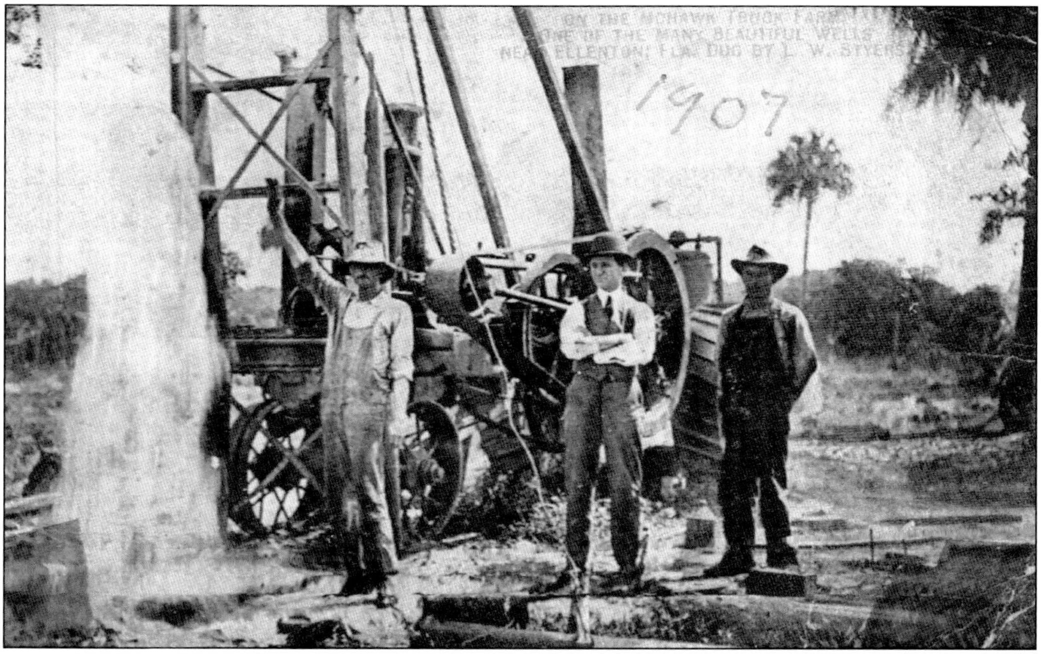

Artesian wells provided water for most of the farms and groves throughout the early 1900s. Three unidentified men stand next to this artesian well, located on the Mohawk Truck Farm near Ellenton. L.W. Stylers dug it in 1907. The prevalence of artesian wells enticed farmers to migrate to Palmetto. (Manatee County Historical Records Library.)

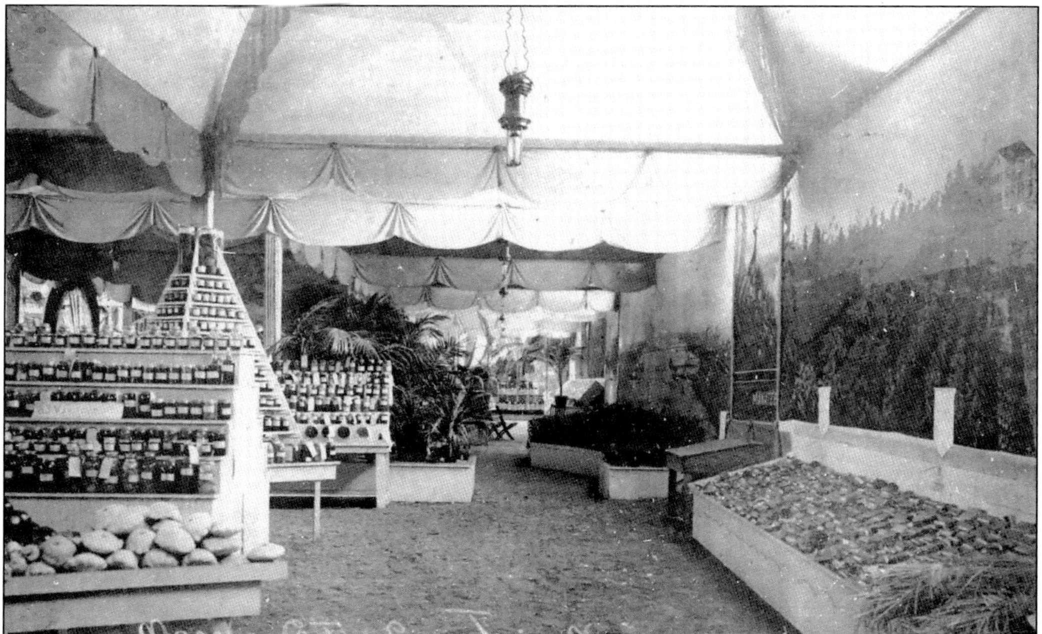

One way farmers were able to advertise their produce to potential customers was by displaying them in a booth at the Florida State Fair. In July 1904, the South Florida Fair Association was formalized, and plans for a new exhibition were proposed. Agricultural organizations across the county jumped at the opportunity to show the extensive variety of farming in the area. (MCHS.)

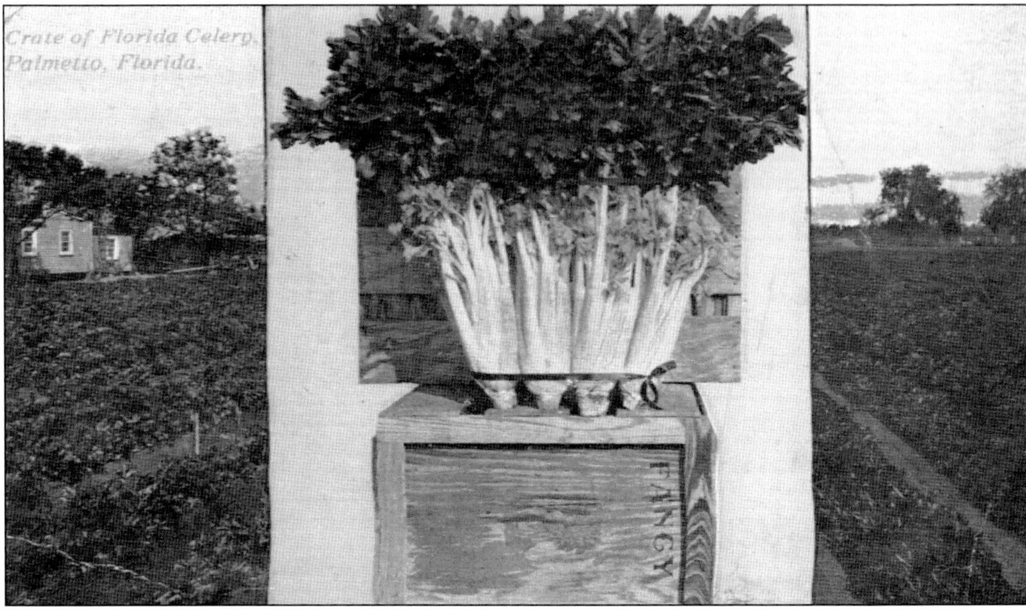

Celery was a popular crop in Manatee County. The plant was staked to ensure straight stalks. In the early 1900s, consumers preferred white celery stalks, so boards or heavy paper were placed along the rows to shield the stalks from the sun and prevent them from turning green. (Manatee County Historical Records Library.)

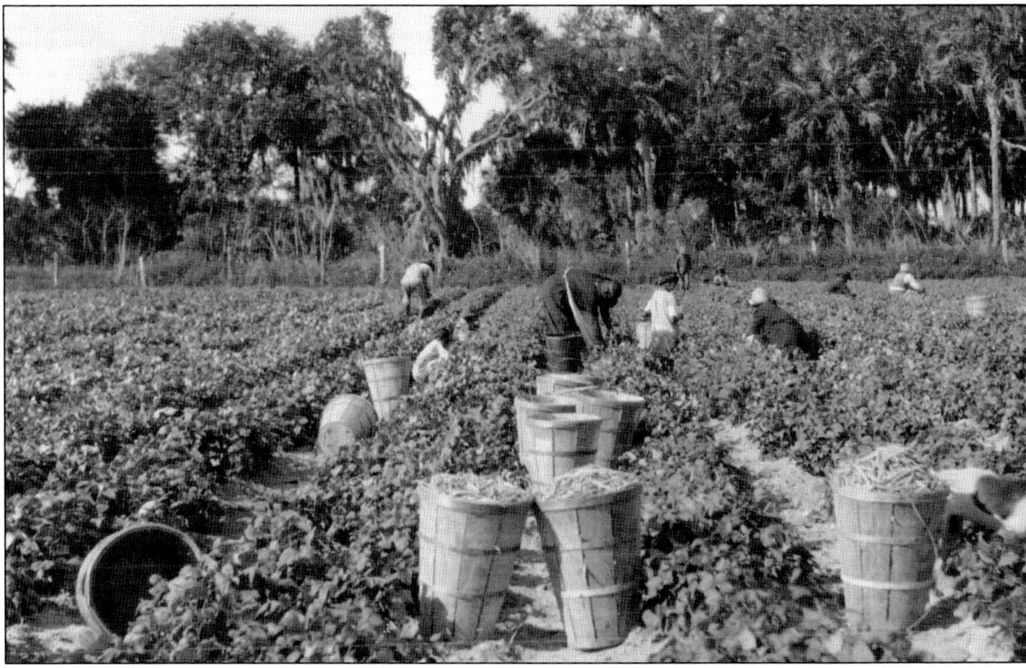

Snap peas are being harvested on Terra Ceia by workers of all ages, including a small child. Peas were among the crops that were picked and shipped "green." By the time they reached the northern markets, they were ready for consumption. Before the first artesian wells were dug in 1895, most vegetable crops were produced on coastal islands like Snead Island and Terra Ceia because they were frost-protected areas. (PHC.)

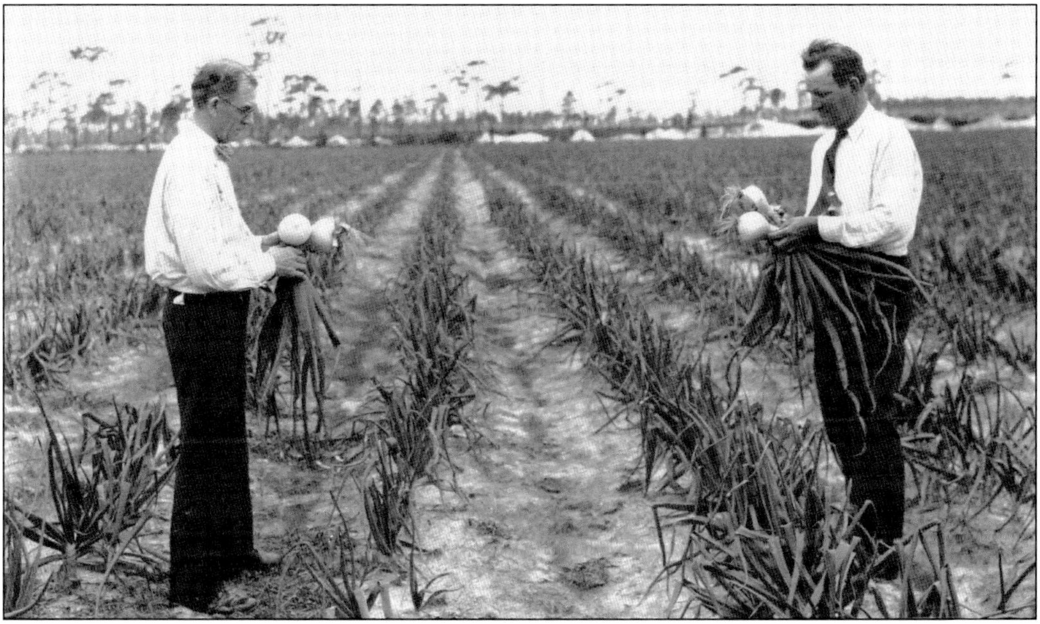

Here, two men inspect their crop of onions. The business of perishable vegetables expanded during World War I. At one time, Big Boston lettuce, peppers, tomatoes, cabbage, eggplant, escarole, onions, celery, and cucumbers were all grown in Palmetto. Most of these crops could be harvested and packed in the field. After a series of disasters in the late 1920s destroyed the crops, most farmers turned to growing tomatoes. (PHC.)

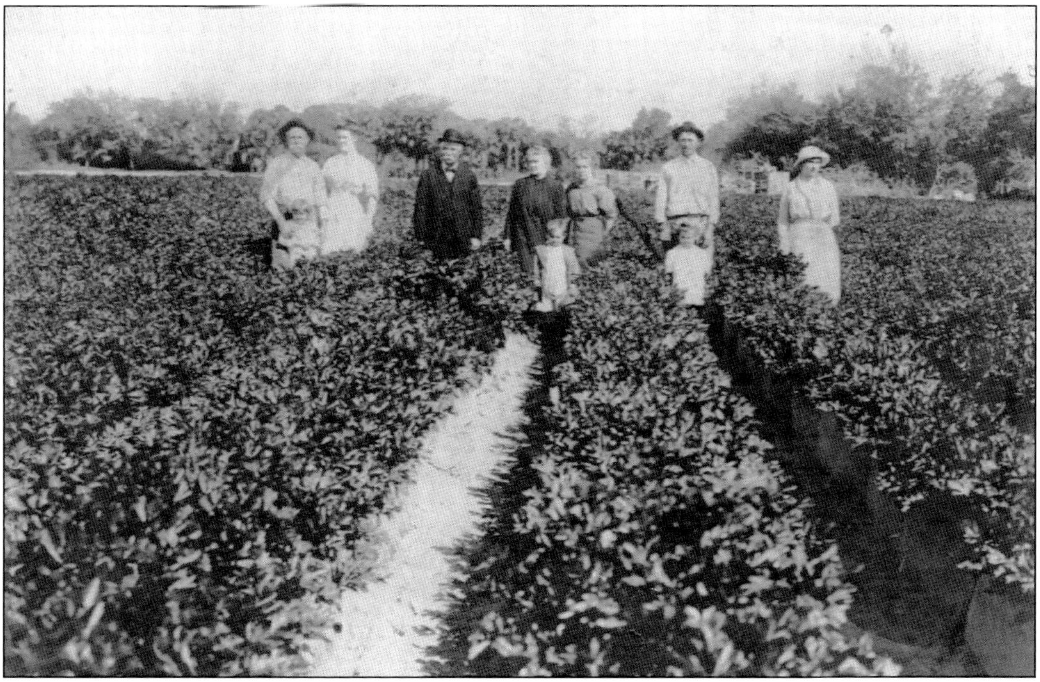

The Moon family is seen at their Frog Creek farm in 1913. Recognizable in the photograph are Lucille Moon, Mr. Flemmings, Hattie Flemmings Moon, John Lanier, Rachel Lanier, James C. Moon, and Selma Moon. (PHC.)

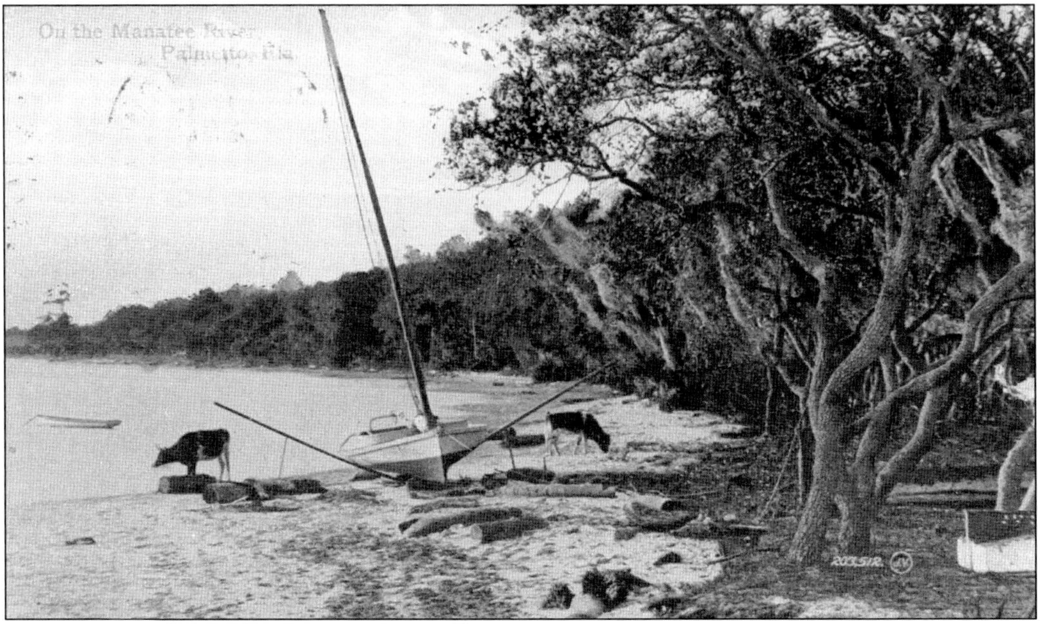

Free-range cattle eat cull fruit dumped into the river from the packinghouse on the Howze Wharf, which was at the end of Thirteenth Avenue in Palmetto. The cattle would come out in the open to feed on discarded tomatoes and citrus. Today, cull fruit remains a favorite treat of cattle, and often packinghouses will dump it by the truckload into populated pastures. (MCHS.)

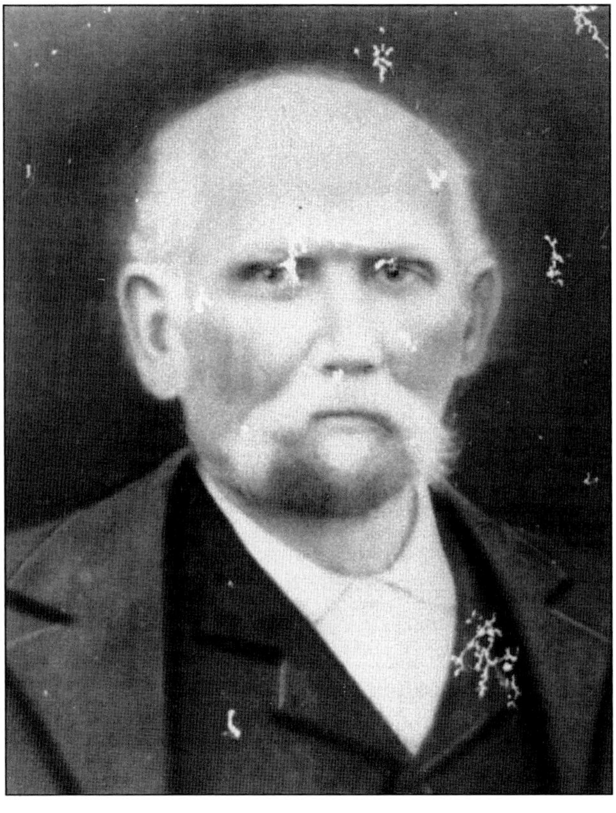

William G. Gillett grew the first tomatoes in Florida in 1875 in the hammocks between Ellenton and Parrish, two towns north of Palmetto. He found that the sandy soil yielded a higher quality of tomatoes per acre. Soon, other growers followed his lead and planted tomatoes. (MCHS.)

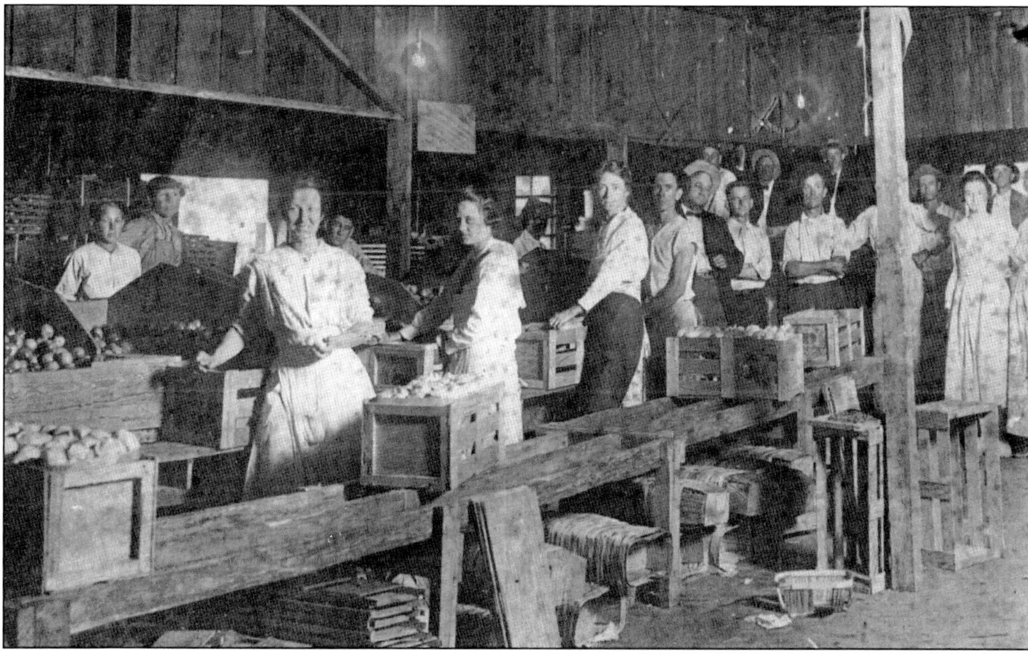

Around 1910, packinghouses started to open around the county, such as one owned by brothers Cleveland and Daniel Marion Courtney that began operation in 1918. Their plant was on Eighth Avenue, with access to the railroad. D.M. married Polina Abbie Koons in 1899, and J.C. married Annie Pedrick in 1911. (PHC.)

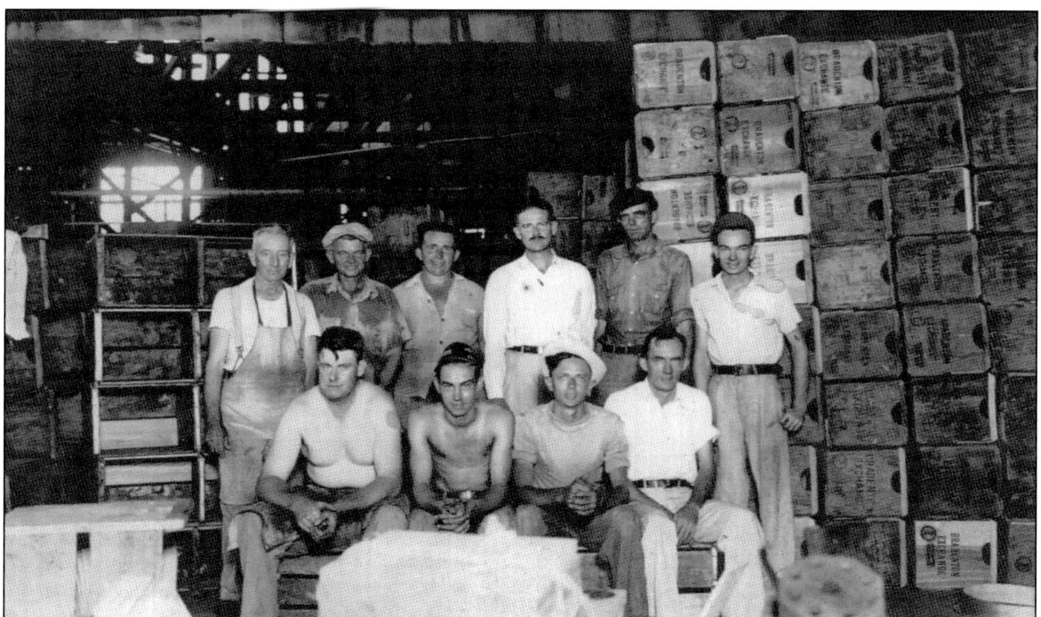

Unidentified workers pose for a picture inside a typical packinghouse. Tomatoes were picked green and placed in ventilated wooden crates, where they would ripen for the 12 days it took to ship produce to New York. J.A. Howze and J.N. Mclean built the first packinghouses at the end of the Palmetto Wharf. (PHC.)

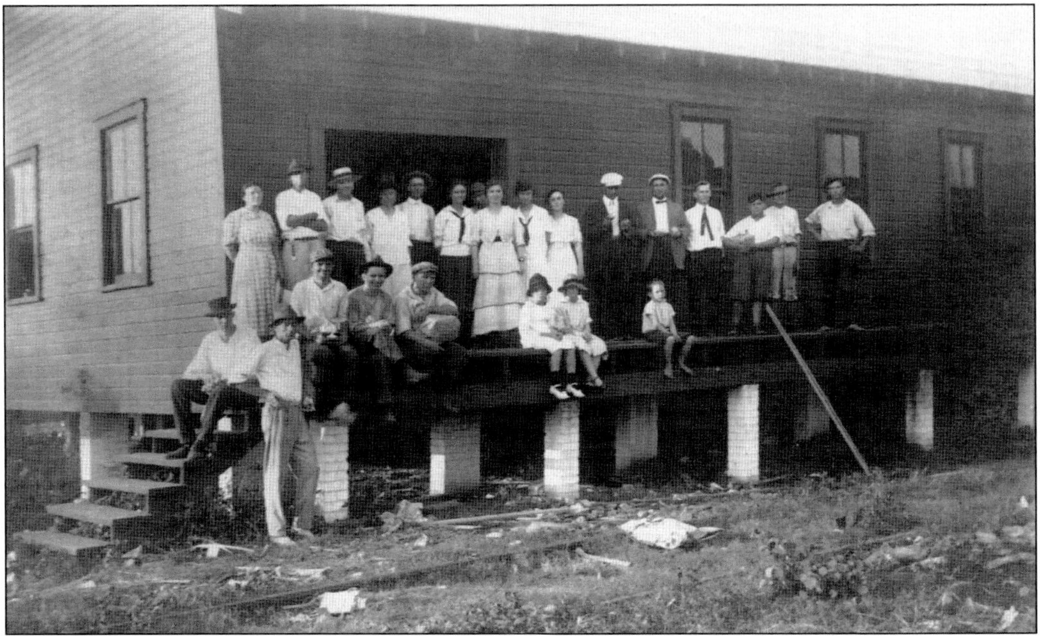

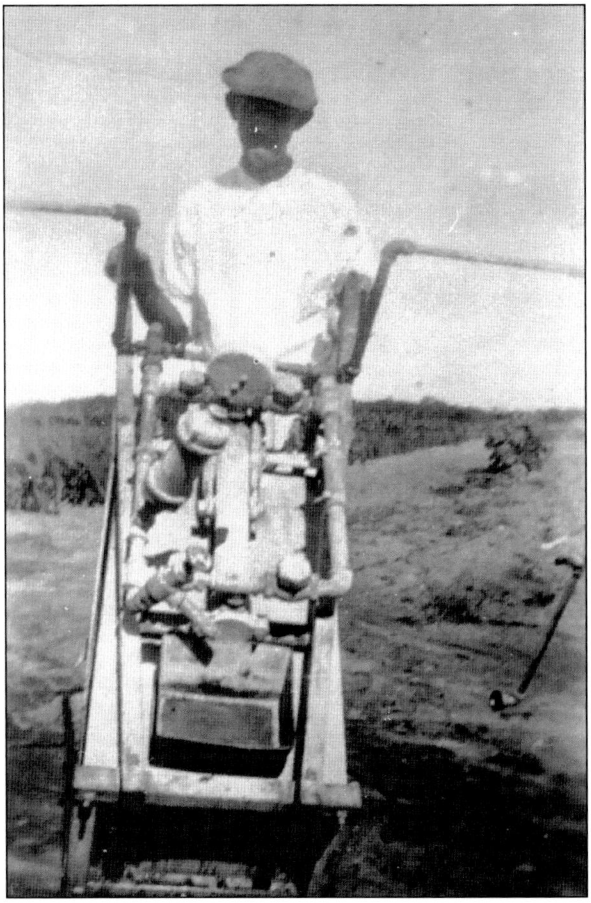

Workers sit outside of an early packinghouse waiting for the train. In 1918, the Manatee Growers Association was organized and remained active until 1960. The association built packinghouses and worked collectively for better marketing, financing, and purchasing of farm supplies and fertilizer. The gradual decline of the group came about when other forms of cooperation, such as state farmers' markets, began to take over. (PHC.)

Grover Cleveland "Shorty" Vowell (pictured) was born in Mississippi in 1885. He came to Ellenton in 1903. In 1929, with the help of Bob Billingsly Whisenant, he invented a horse-drawn sprayer. L.A. Rawls did the welding at the Harllee and Harrison garage in Palmetto. The sprayer was patented and manufactured at the Vowell Sprayer Company in Ellenton. (MCHS.)

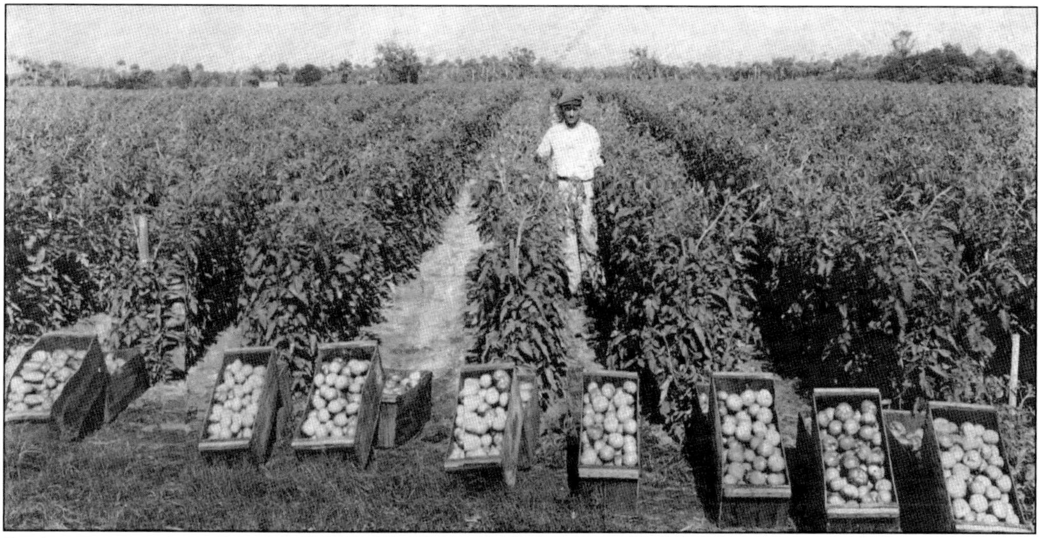

Staking tomatoes was a technique used in Manatee and Hillsborough Counties for several years before being adopted throughout the United States. Shorty Vowell is credited as the first grower to use stakes, which were made of bamboo or palm branches. This small innovation was an improvement, because it made harvesting easier and more efficient. Ground-grown plants had a delayed maturity rate and were more susceptible to disease and decay. (PHC.)

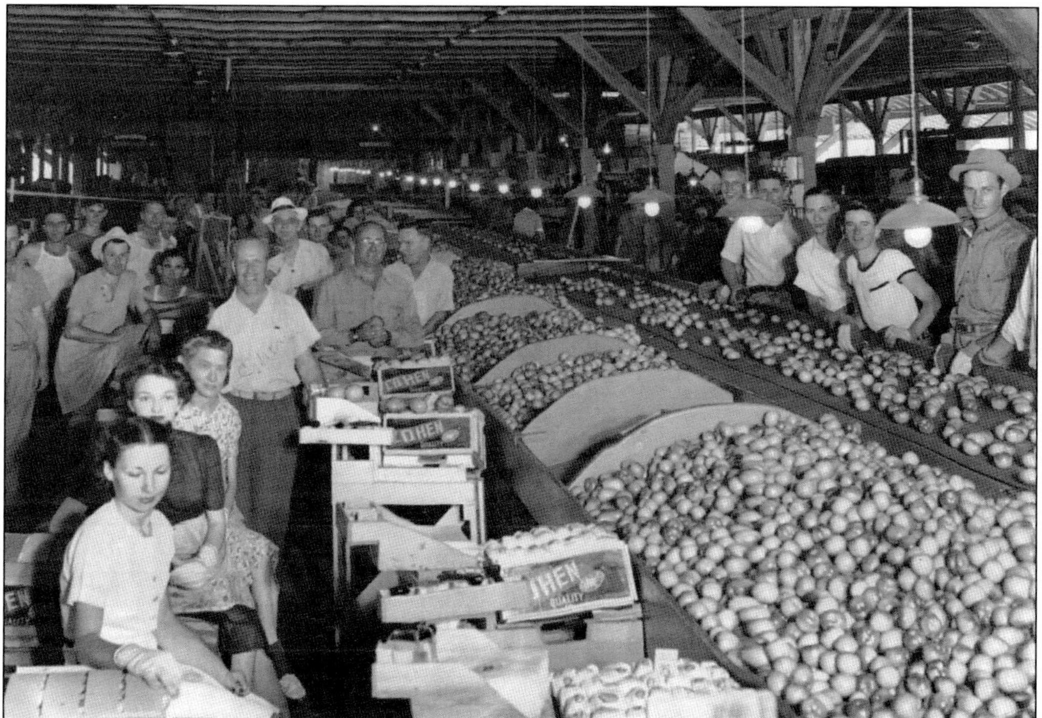

John Neil McClure Sr. arrived in Palmetto in 1920. He purchased 10 acres of land just north of Palmetto and planted lettuce, tomatoes, and a variety of other vegetables. Pictured here is McClure's original packinghouse on Seventh Street and Sixth Avenue along the Atlantic Coastline Railroad. In 1962, the Atlantic Coastline Depot became the West Coast Tomato Building. (PHC.)

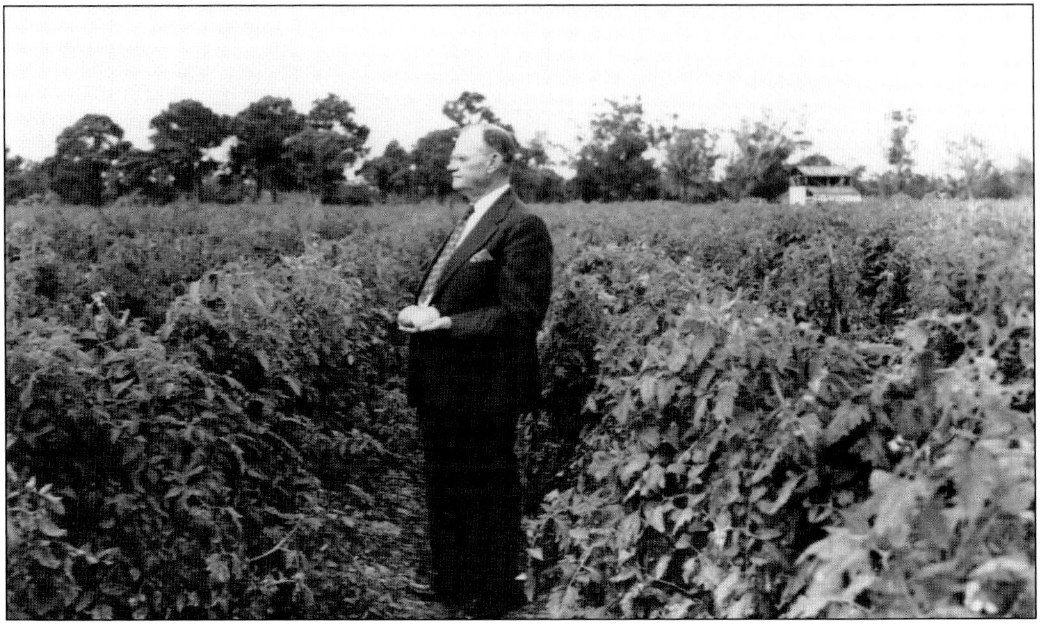

John Pope Harllee was an entrepreneur who opened the first Ford garage in Manatee County on Fifth Street in Palmetto; however, Harllee is most noted for his tomato farming. In 1916, he started Harllee and Harrison Packinghouse on Canal Road. It was destroyed by fire in 1936. He then opened Harllee Farms Packinghouse on Tenth Street, which today is one of the nation's largest tomato businesses in operation. (PHC.)

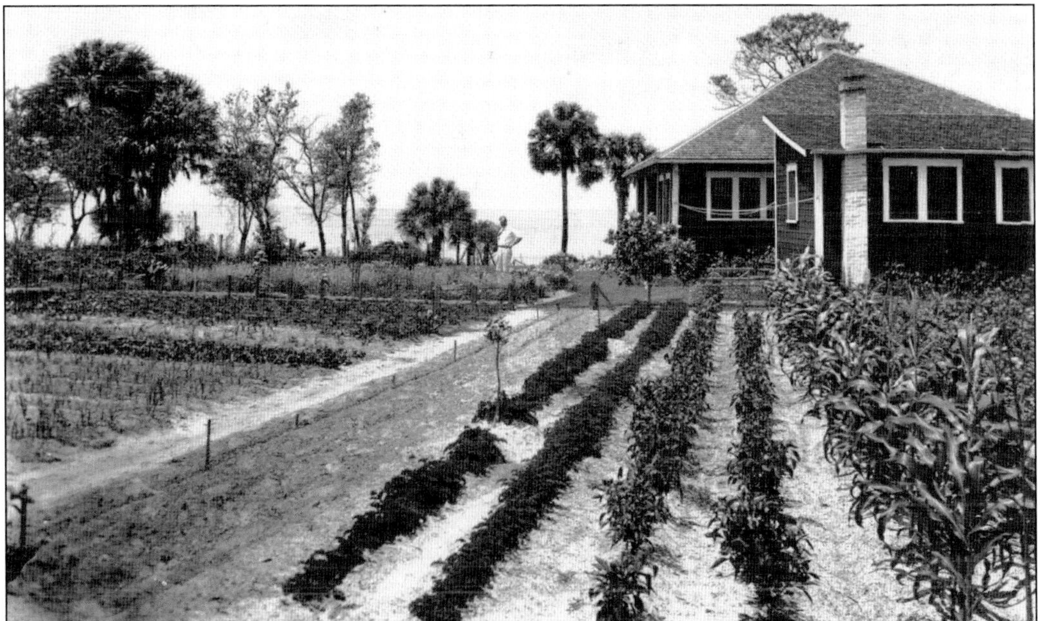

Mead Smith stands in his home garden, where he grew a variety of vegetables. Small plots like Smith's were common for market gardening, which is the cultivation of crops on suburban land for the supply of nearby cities. Home gardeners could sell their crops at the State Farmers' Market, which was built in 1936 on Eighth Avenue where the Palmetto Plaza stands today. (PHC.)

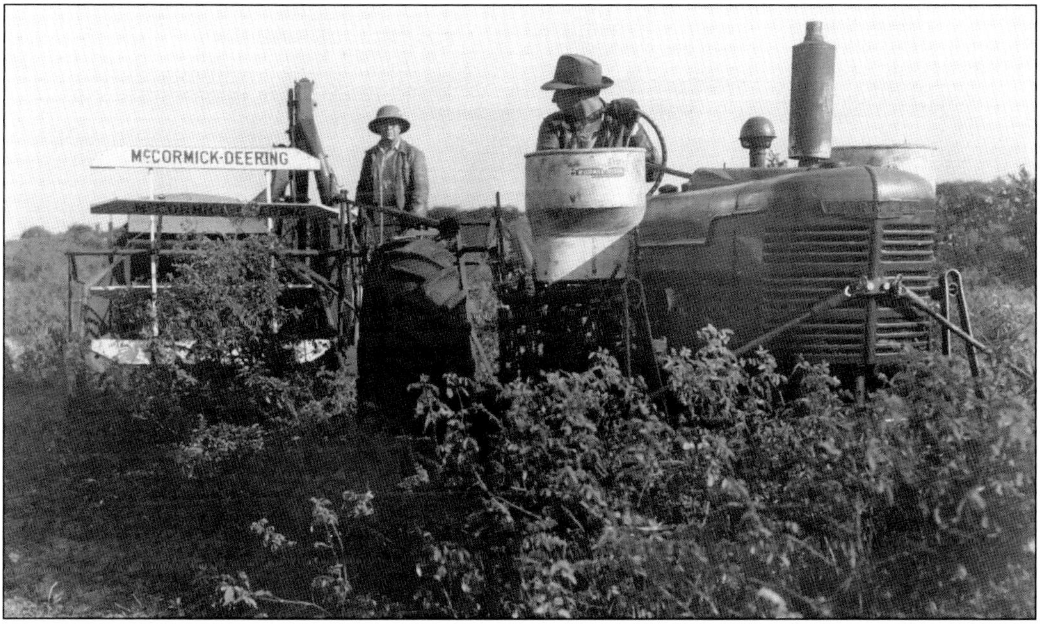

The first two decades of the 20th century have been called the golden age of American agriculture. New machinery offered by manufacturers seemed to spring up in every city, town, and village in the agricultural heartland. McCormick Deering Tractors were manufactured in Chicago, but they improved farming all over the country. Here, Leland Jefferies uses his tractor to plow fields. (PHC.)

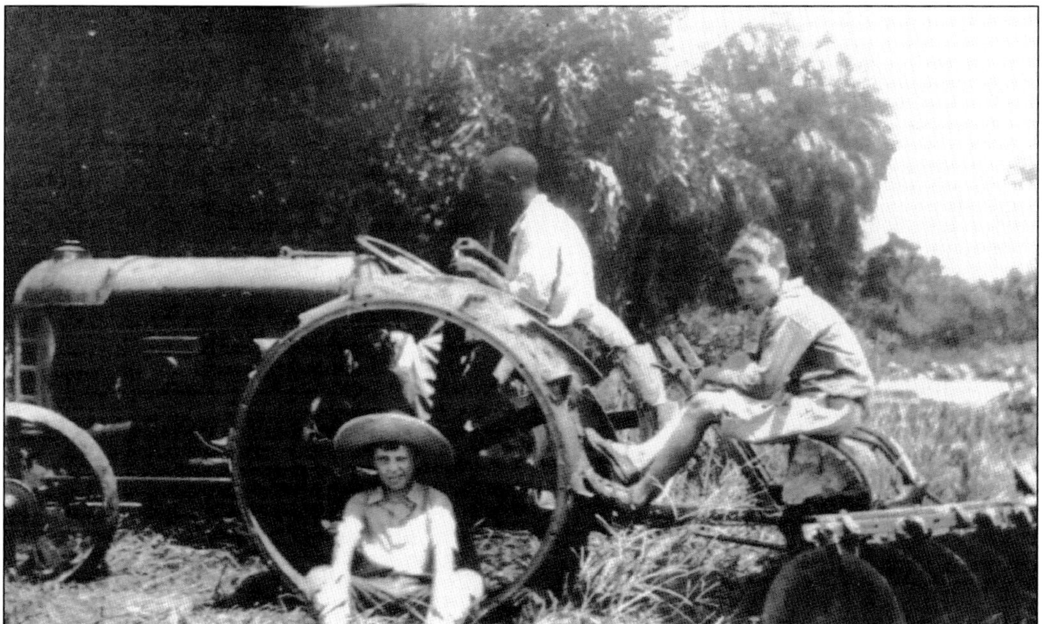

Jessie Butler is shown at the wheel of a Ford tractor in 1918, George Kermode is riding on the disc, and Rad Kermode is inside the wheel. Henry Ford incorporated his new Henry Ford & Son Tractor Company in 1917. Ford tested his competitor's tractors on his family farm. Prominent Palmetto farmers were some of the first to purchase the new model. (PHC.)

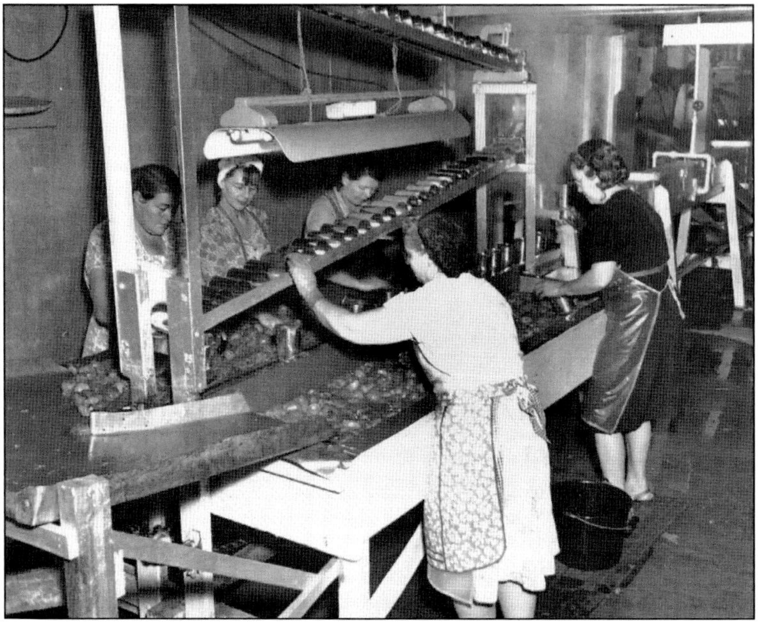

John Greenlaw founded Palmetto Canning in 1927. His plant only produced guava jelly. After World War II, John Greenlaw Jr. assumed control and diversified the operation into the canning of tomatoes and other produce. Around 1950, after a downturn in the tomato industry, Greenlaw decided to return the company to its core business of guava jelly. Today, the company packages all its products in glass or plastic. (PHC.)

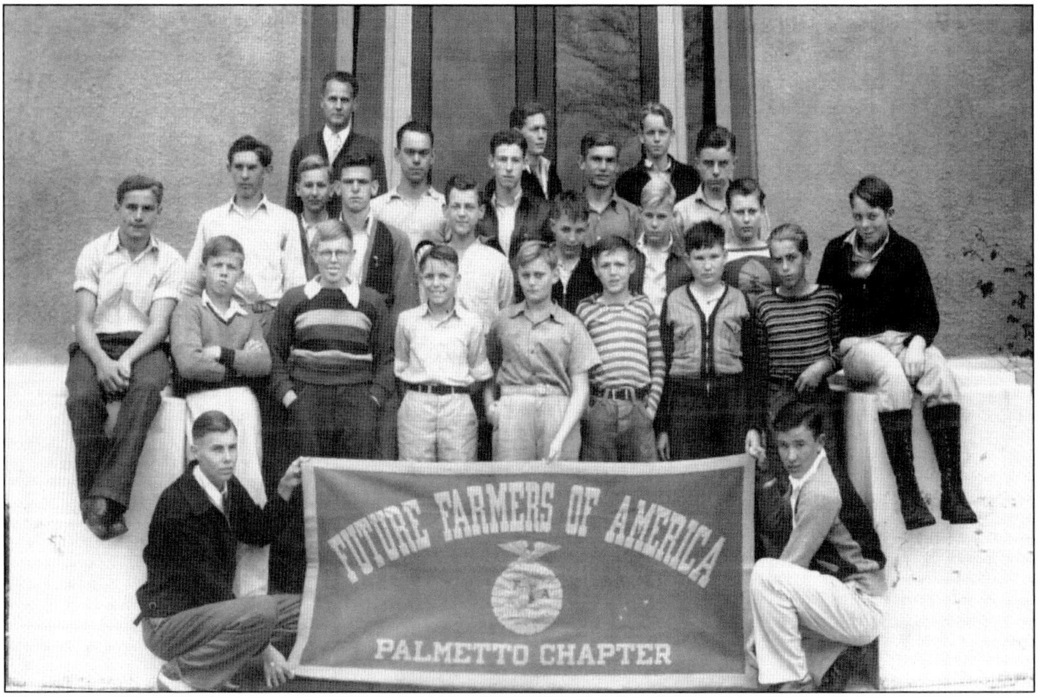

The Future Farmers of America Palmetto chapter was formed in 1929. Every year, regional chapters competed for Future Farmers Day at the state fair. Palmetto consistently won for its livestock and agricultural exhibits. In 1942, the boys constructed their own headquarters—even plumbing the premises themselves. The building is on the corner of Seventh Street and Tenth Avenue. Vice Pres. Billy Parks and Pres. Douglas Turner hold the banner. (PHC.)

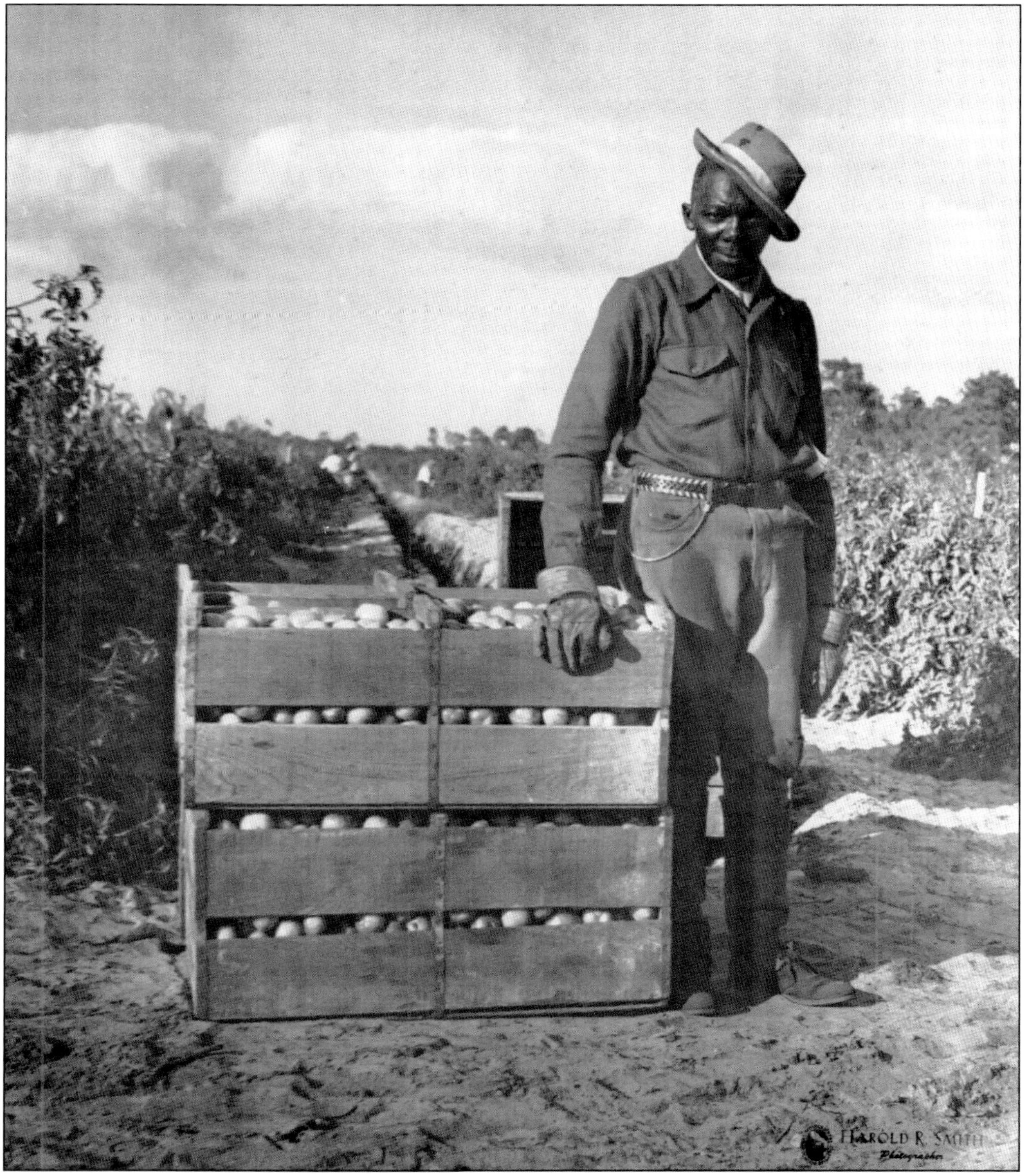

Historically, a small number of African Americans farmed successfully on their small acreages, selling their produce through the State Farmers' Market in Palmetto. However, in 1949, the market was closed down, forcing them to truck their wares to Tampa. In 1957, students taking lessons from Emett McCray, a Lincoln High School agriculture teacher, developed a co-op called Mana Hill Packing Company for small-scale African American farmers, since major packinghouses refused to buy their produce; however, the organization could not get a start-up loan. McCray solicited assistance from an Alabama group that worked with minority farmers, and in 1972 the co-op had its first season, operating out of an abandoned packinghouse scheduled to be razed. At its peak, Mana Hill employed 150 workers, depending on mostly Mexican and Haitian migrants for harvesting. Southern Cooperative Development Fund planned to partner with Mana Hill and purchase 3,600 acres for a small-farm project, but the county commission refused to approve the deal. An unidentified worker leans against tomato crates. (PHC.)

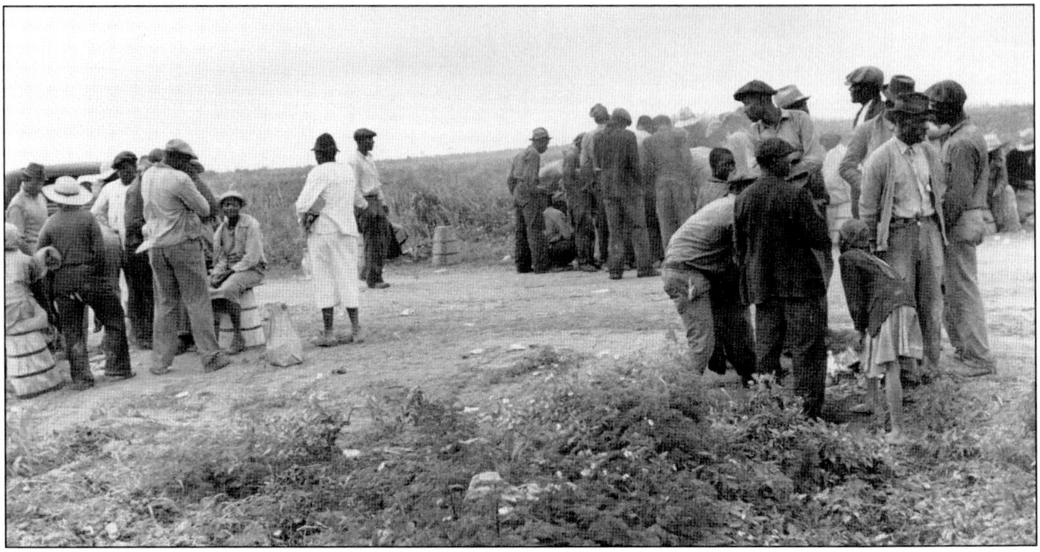

During the 1930s, crew leaders would truck African Americans and migrant workers north to pick crops for harvest season. These men employed many residents of Rubonia to work the fields. It was a profitable business for crew leaders and white growers, but field workers were often exploited, earning very little pay. Workers are seen waiting to collect their pay in this 1930s photograph. (SLAF.)

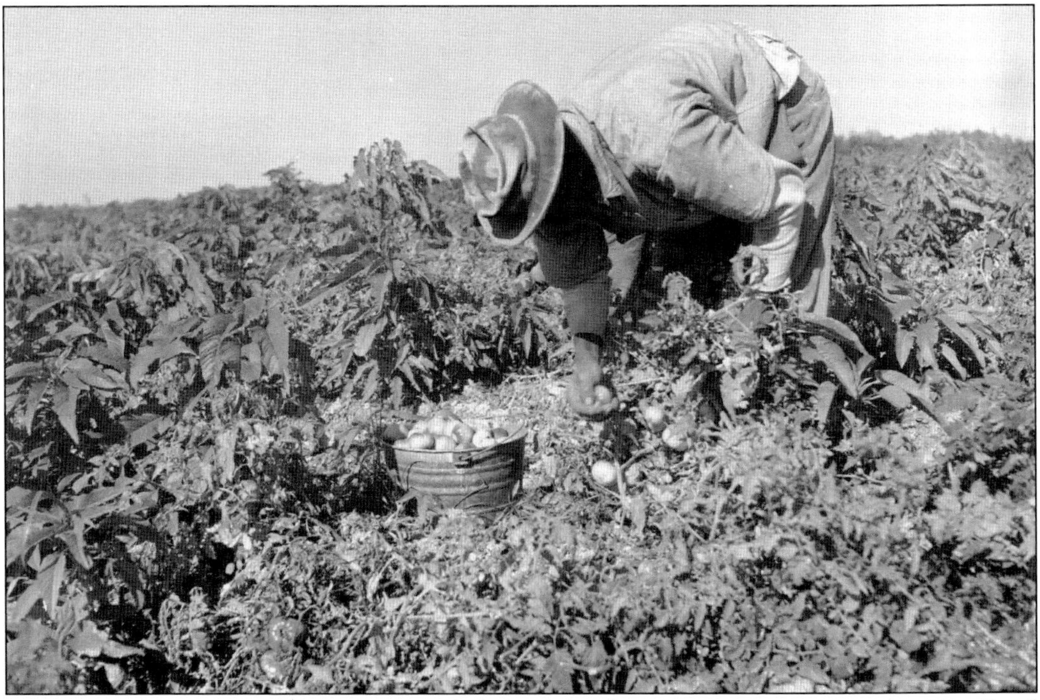

During World War II, a labor shortage led the federal government to set up the War Food Administration, which brought in farmworkers from the West Indies. They were housed in the exhibit buildings of the Manatee County Fairgrounds, which is now occupied by the Palmetto Trailer Park. (SLAF.)

Four

A Community of Commerce

From the beginning, Palmetto's climate and location contributed to bountiful harvests and spurred the development of commercial agriculture. Whenever early settlers had a surplus of produce or livestock, they bartered with other families to trade in goods and services. Eventually, plantation owners from Tallahassee traveled south to capitalize on the area's potential for cash crops. As early as the 1850s, sugar grown in the vicinity of Palmetto was exported on sailing vessels; the Cuban cattle trade soon followed. As agriculture expanded and a string of ships carried cargo to the outside world, Palmetto residents envisioned their slice of riches.

Julia "Madame Joe" Atzeroth was Palmetto's earliest non-agricultural entrepreneur. After a few years of running a successful beer and cake shop in Tampa, she purchased an ideal piece of property on the Manatee River at present-day Eighth Avenue and Fifth Street. In 1851, she opened the town's first store in a building that had additional lodging for river travelers.

With the construction in the 1870s of commercial wharfs to handle the heavy traffic of cargo ships, businesses and shops sprouted along the riverfront. At first, the mercantile trade was comprised of mom-and-pop shops, many of which operated from the docks. Palmetto's economy depended on the Manatee River, and when the community created a chamber of commerce in 1895, it was called the Manatee River Board of Trade. All but one of its members was from Palmetto.

The population grew quickly, and the offspring of the settlers sought their own trades and businesses. Soon, the city abounded with blacksmiths, boat builders, hotel proprietors, and shopkeepers.

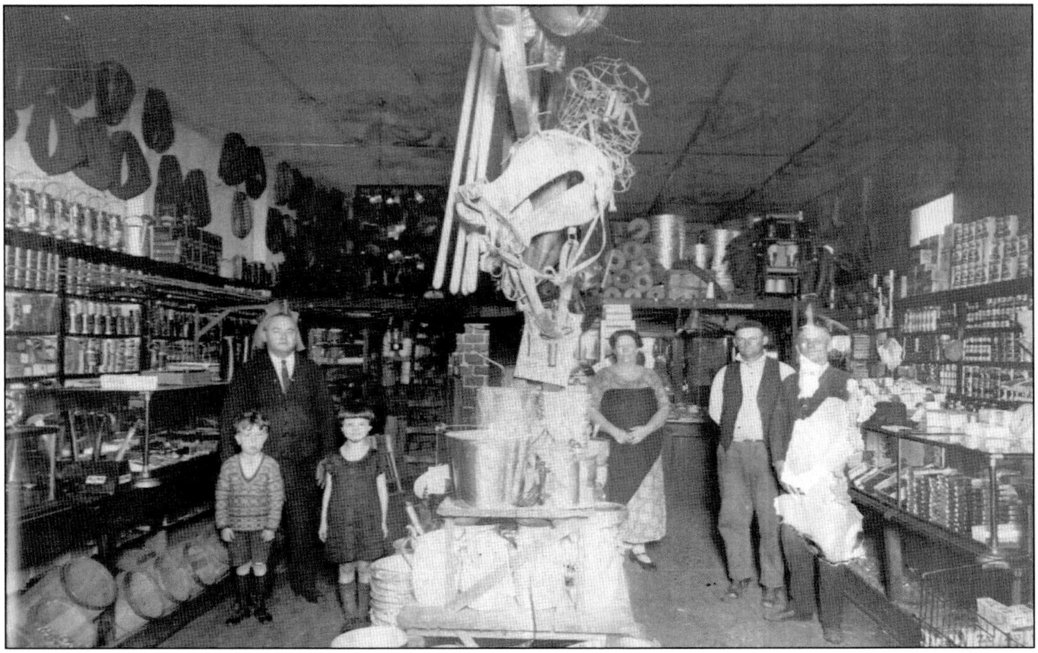

S.S. Lamb established Palmetto's first dry goods store shortly after the town was named in 1868. Reputed to be the largest and best-equipped grocery in the county, it stocked a variety of canned goods the local residents valued, especially in the agricultural off-season. (PHC.)

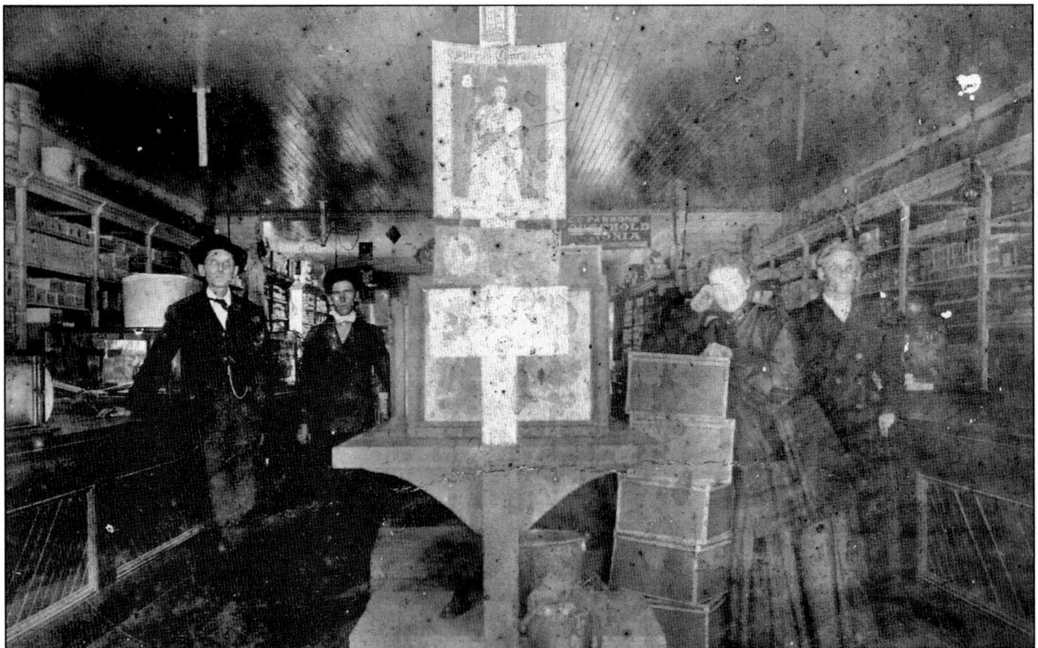

James K. Parrish and his wife moved to Palmetto during a period of rapid growth. In 1885, they opened the Old Reliable, pictured here, the second grocery store in town. A rivalry developed between Parrish and S.S. Lamb, who did not initially welcome the competition. But when Parrish's son Walter Eugene married Lamb's daughter Nellie, the two settled their differences. (SLAF.)

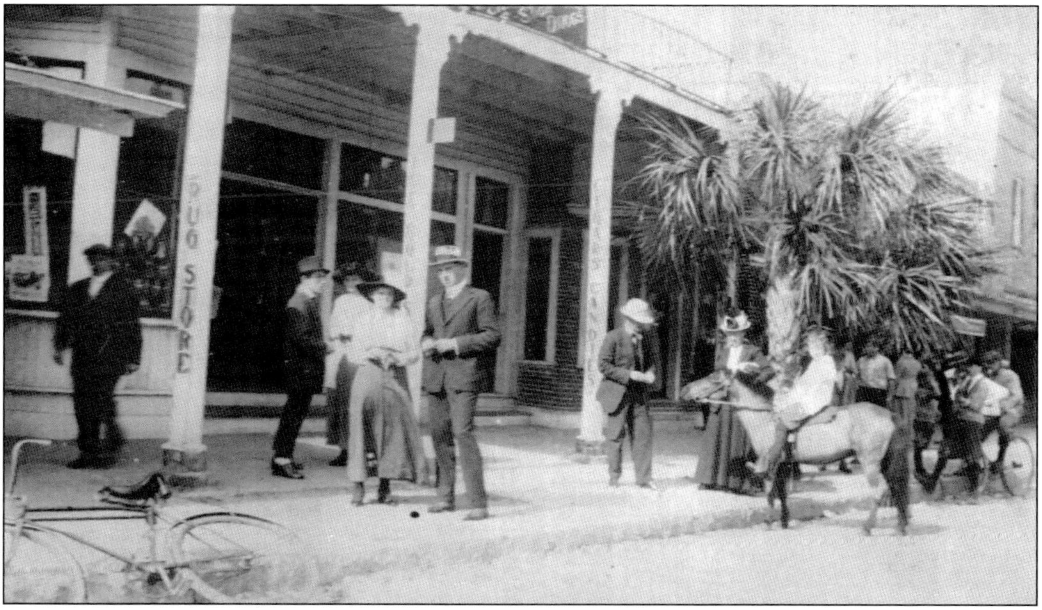

George T. Dickey emigrated from Scotland and came straight to Palmetto in 1888. He married Bessie Harrison, the daughter of Dr. M.B. Harrison, owner of the Palmetto Drug Company. Dickey took over the pharmacy, and it came to be known as Dickey's Drugs. Here, Ruth Davis rides her pony to pick up a prescription. (PHC.)

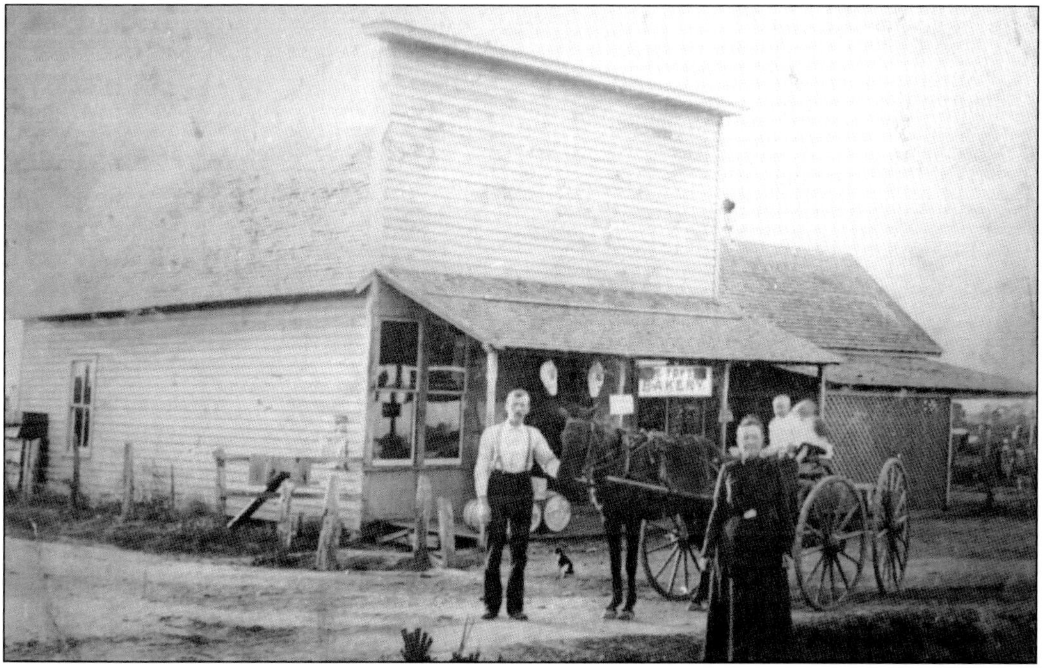

Elizabeth Toft, her son Walter H., and his son Walter H. Jr., along with daughter Ruth Ann and their horse Sam, pose in front of their business. Originally from England, the Tofts moved to Palmetto in 1894 and opened a profitable restaurant and bakery in their home on Twelfth Avenue. The aroma of Elizabeth's fresh baked bread beckoned traveling salesman. The business relocated to Eighth Avenue and Seventh Street in 1898. (PHC.)

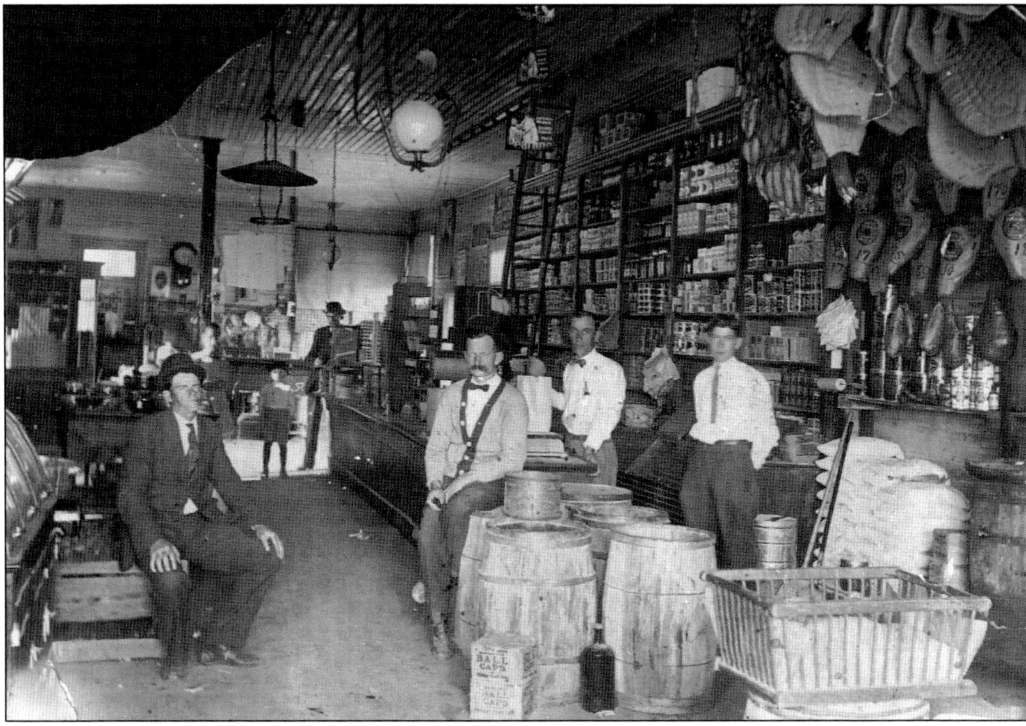

The interior of the C.P. Fuller Store on Central Avenue and Palmetto Road in Ellenton (present-day US 301) is pictured here. In the 1890s, it was the largest general store south of Tampa, carrying everything from groceries to coffins. It even had an elevator. J.R. Wood is on a crate, an unidentified boy stands next to Murray Ballinger, Earnest Fischer is seated on a barrel, and L.S. Hardin and Will Fischer are behind the counter. (PHC.)

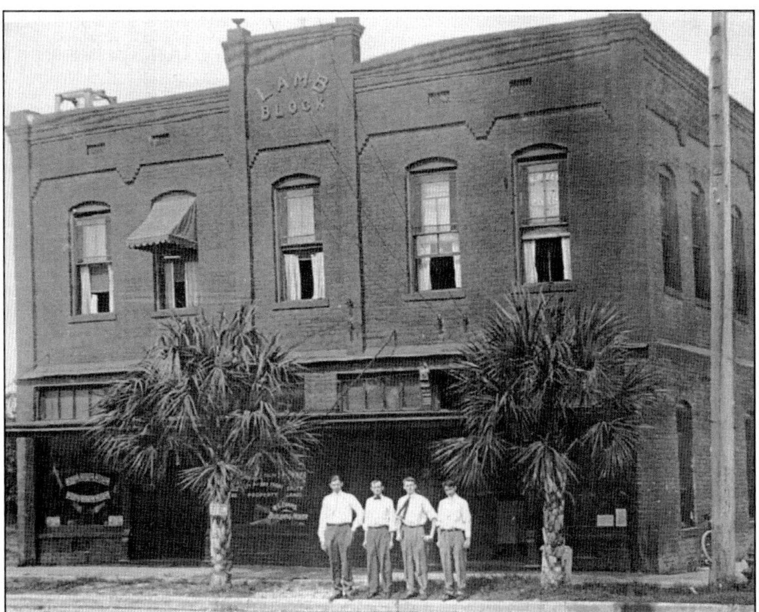

S.S. Lamb constructed Palmetto's first brick building to house the Manatee County State Bank. Later, a number of different businesses operated within its sturdy walls. The century-old structure still stands on Main Street (now Tenth Avenue) in Palmetto. (PHC.)

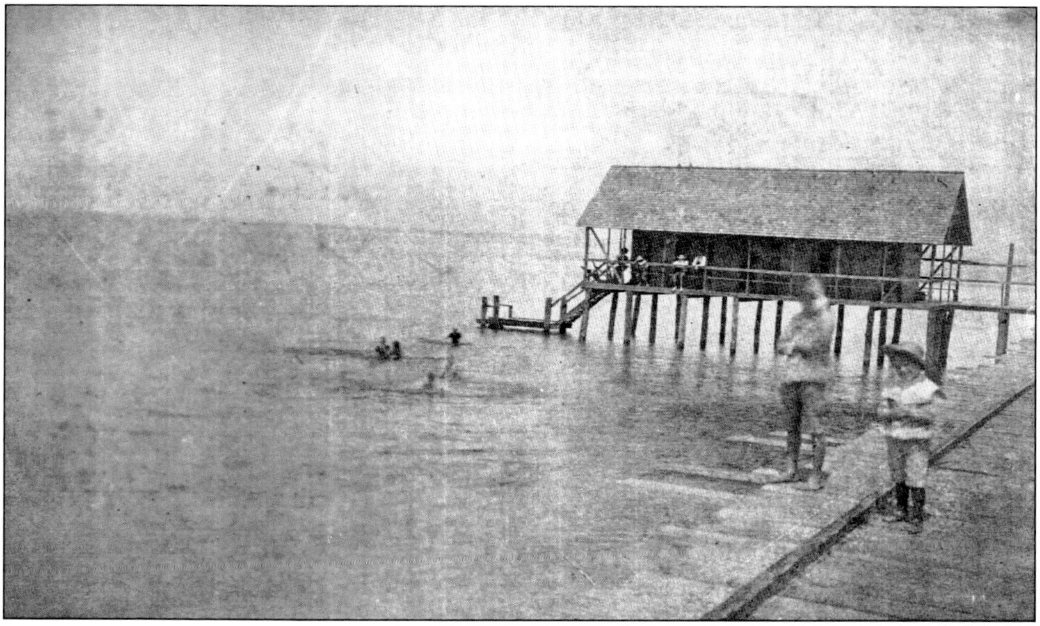

The Palmetto Inn was attached to Hendrix's Wharf by a small walkway. Ship passengers and crews ate at the restaurant, operated by Waddy A. Bacon. The inn also served as a seafood market. The inn was moved to City Dock in 1897, a popular spot for swimmers. (PHC.)

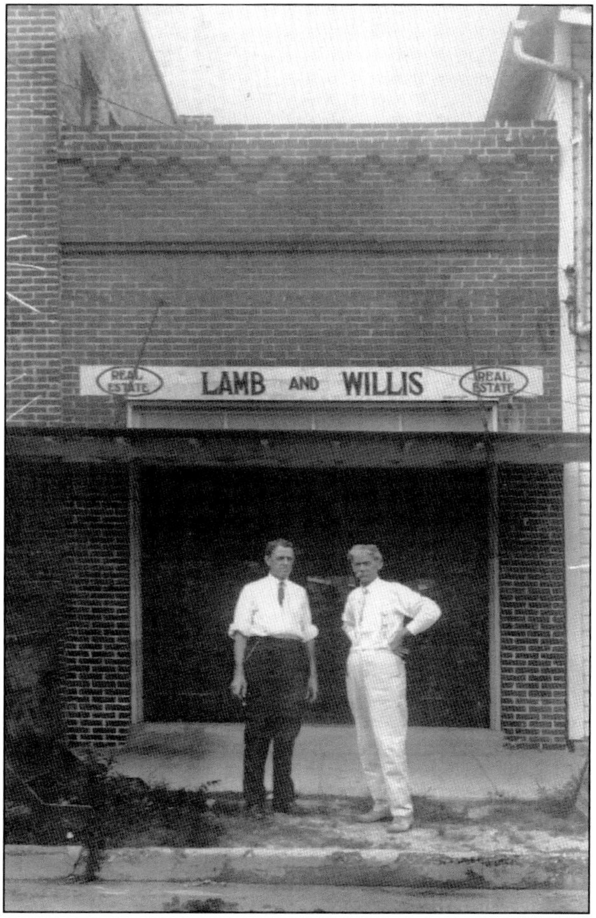

Julius A. Lamb was another of Palmetto's entrepreneurs. He operated a livery stable in partnership with W.T. Chapman around 1893. Then, in 1897, he built City Dock with his father S.S. Lamb and J.A. Lamb. In 1908, Julius tried his hand at operating a dry goods store. By 1916, he joined Lamb and Willis, a real estate firm in the Lamb-Willis Building on Palmetto's Main Street. (PHC.)

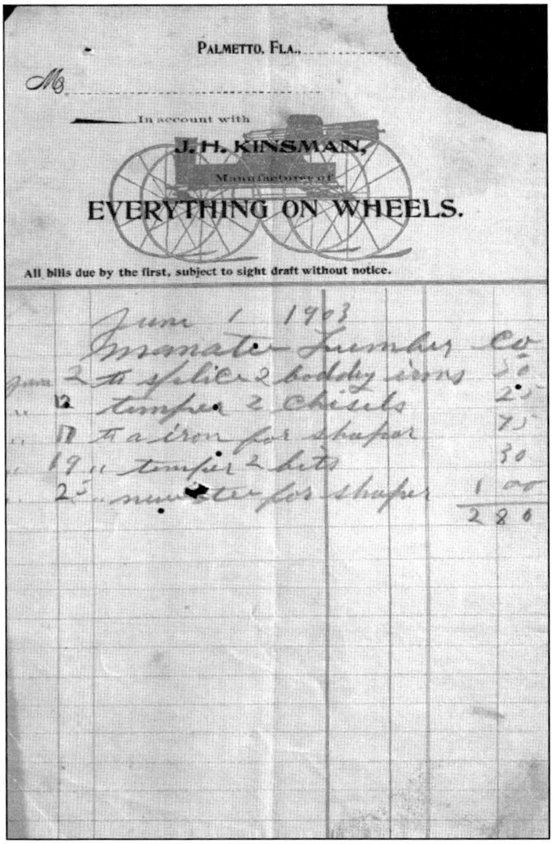
PALMETTO, FLA., ...

M ....

In account with

J. H. KINSMAN,

Manufacturer of

EVERYTHING ON WHEELS.

All bills due by the first, subject to sight draft without notice.

June 1 1903

Manatee Lumber Co

June 2 to splice 2 body irons 50

12 temper 2 chisels 25

17 to a iron for shaper 75

19 temper 2 bits 30

25 [illegible] [illegible] for shaper 1 00

2 80

J.H. Kinsman, Palmetto's first blacksmith and carriage maker, had his shop by the river between Eighth and Ninth Avenues. Among artifacts from his business is this invoice, totaling $2.80 that featured Kinsman's slogan, "Everything on Wheels." A note on the invoice reads, "All bills are due on the first of the month, subject to slight draft without notice." (Courtesy Nick Baden.)

J.H. Kinsman also ran the Kinsman Building, an apartment house near the old Palmetto City Dock. A fire later destroyed the wood structure. In the 1990s, the Regatta Point Marina and shopping area occupied the site. (PHC.)

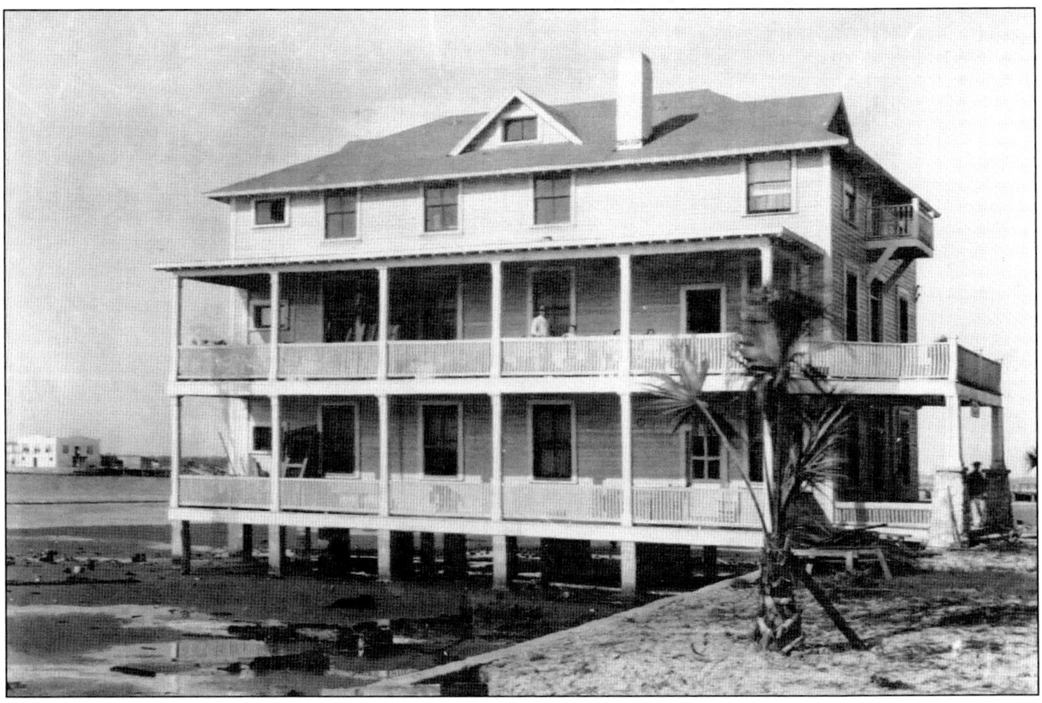

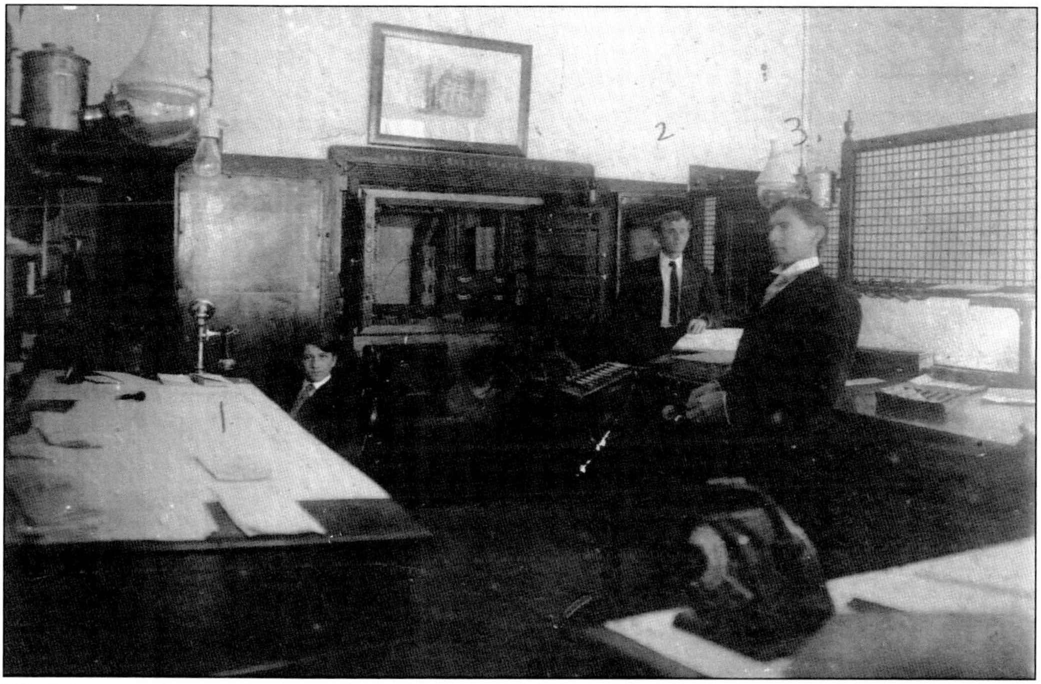

The Manatee County State Bank opened in 1902 on Main Street; it was the first bank in Manatee County. Otaray Waller, J.A. Howze, and Asa Lamb are seen at work, with the bank's open safe in the background. The building now is home to Norma Rae's restaurant. (PHC.)

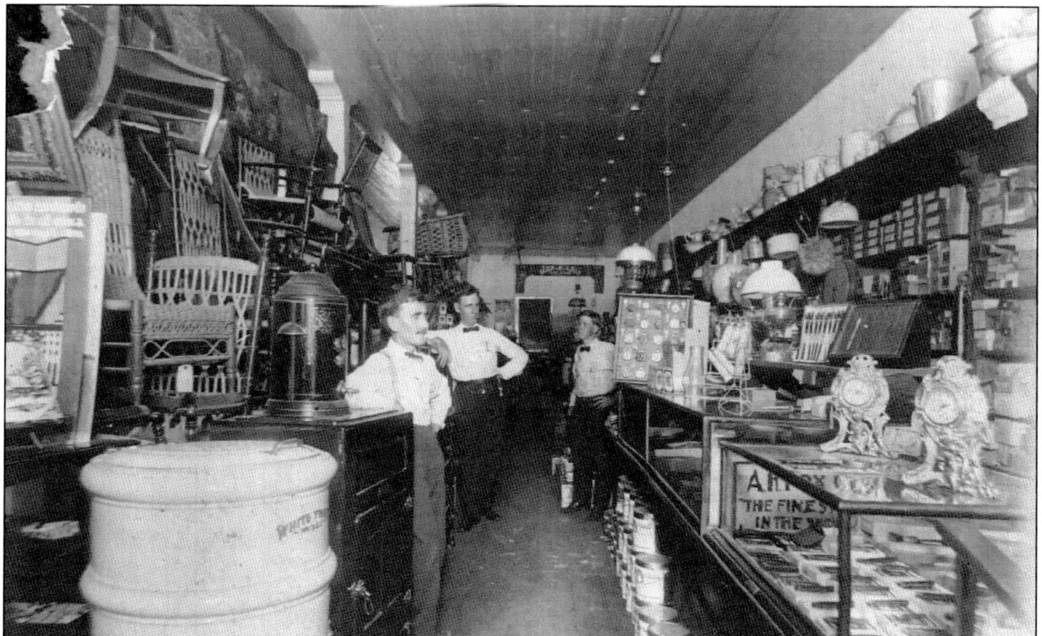

J.B. Owens and J.W. Nettles owned this hardware store where they sold home goods, such as chairs, lamps, and cooking utensils. In this 1910 photograph, Owens assists Mr. Grimes, a drummer, while employee James Hendrix scans cluttered shelves. The glass cases held items like pocketknives and hand tools. (PHC.)

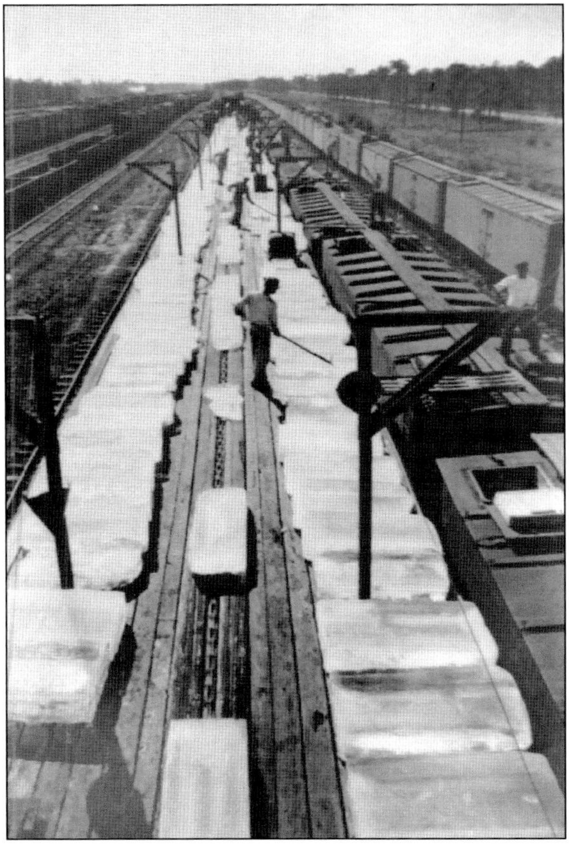

Around 1900, ice was one of the area's largest exports. The first ice factory in Manatee County was the Palmetto Ice and Power Company on the east side of Thirteenth Avenue. The Tampa Ice Company opened in Palmetto in 1909, and the Seaboard Air Line Railroad built two tracks to accommodate its operations. The plant, located on Tenth Street, could produce 60 tons a day. (PHC.)

Charles Porter's fish market was located on Eighth Avenue in Palmetto. He is pictured standing out front with his small daughter in the 1920s. Porter's sold fresh fish from local commercial fishing companies. The building was constructed of metal scraps. (PHC.)

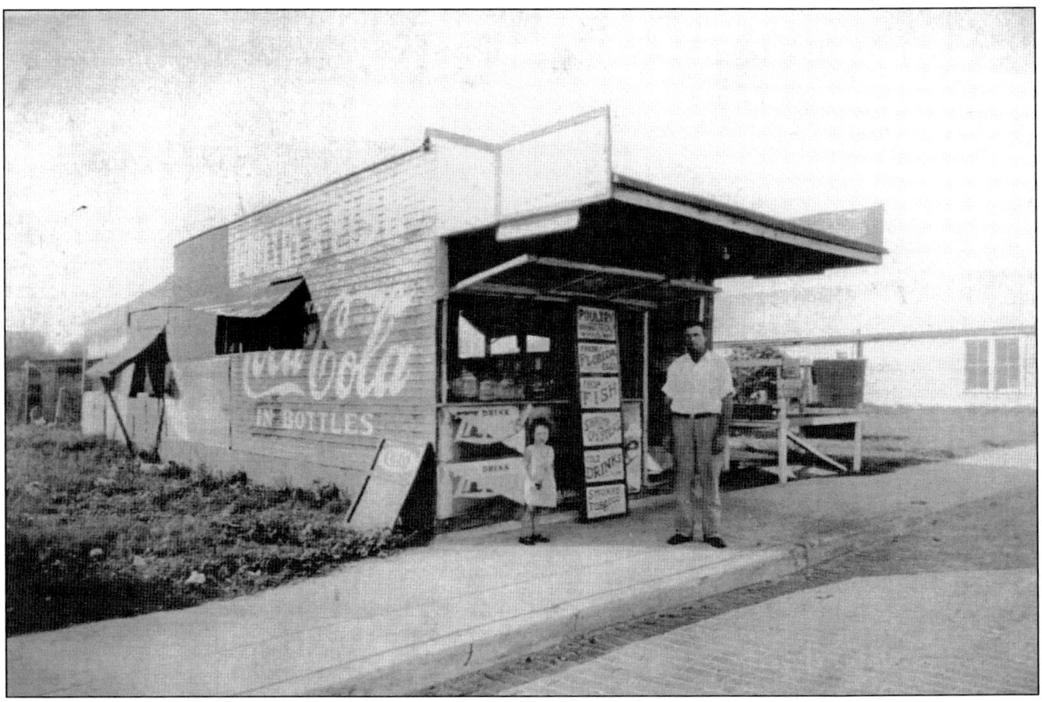

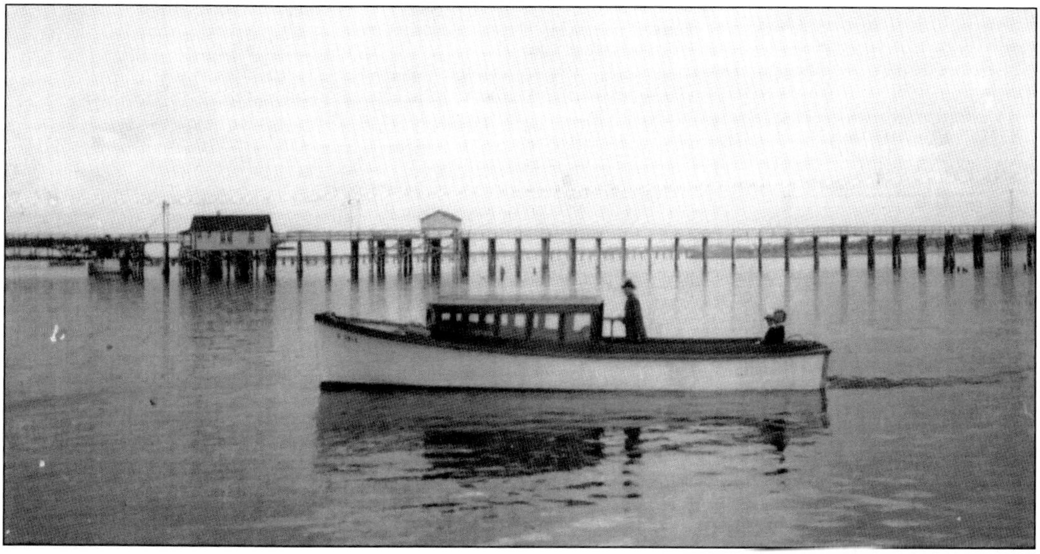

Ed Pillsbury opened the Snead Island Boatworks in 1907, said to be the largest and most profitable boat building and fishing business from Fort Myers to St. Petersburg. Above, one of his boats motors past Victory Bridge. With the business a success, Sam Andress bought the boat works for a period of time but then sold it back to Ed, who owned it until his death. Ed's sons inherited the business, and they sold it to E.E. Bishop, who used it to harbor his private yacht. Bishop hired Andrew Alderman as his captain. After Bishop's death, Jim Alderman, Andrew's son, purchased the boat works; the Aldermans still own and operate the business today. Below is an aerial shot of Snead Island Boatworks by photographer Harold Smith in the 1940s or 1950s. (PHC.)

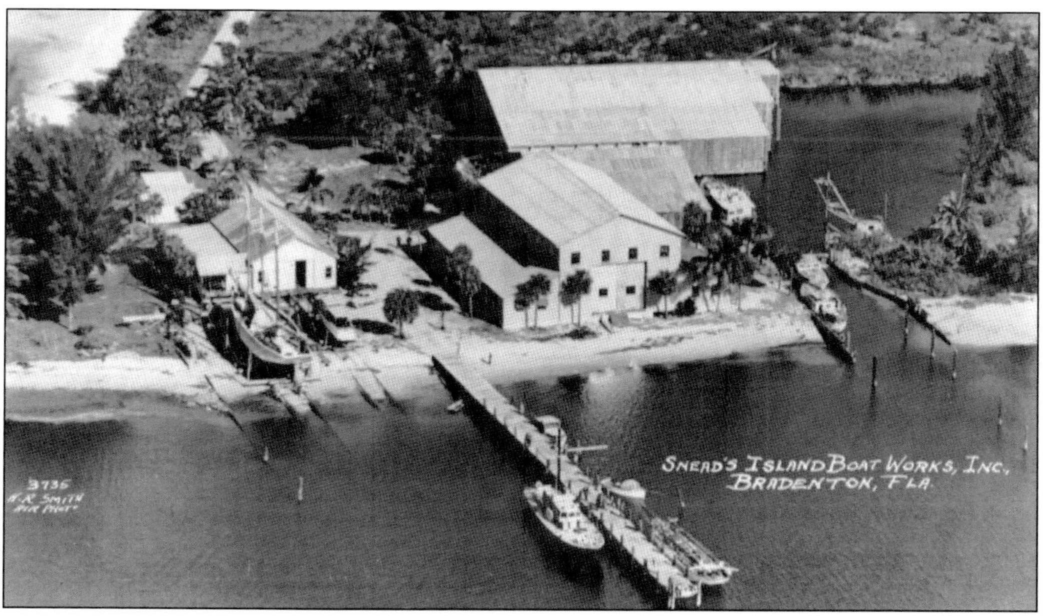

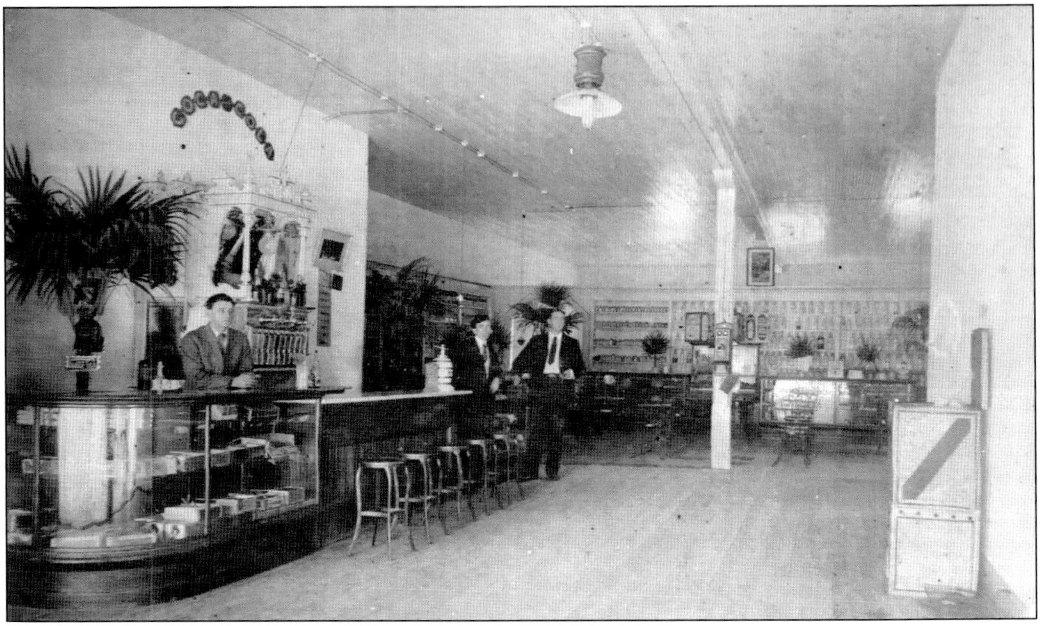

The Red Cross Pharmacy opened in 1908. The cigar bar sold real Cuban cigars imported from Havana—a favorite among Palmetto men. The soda fountain offered Coca-Cola and treats like ice cream and candy. This photograph was taken on opening day, and the shop was still awaiting delivery of tables and chairs. The people in the photograph are unidentified. (PHC.)

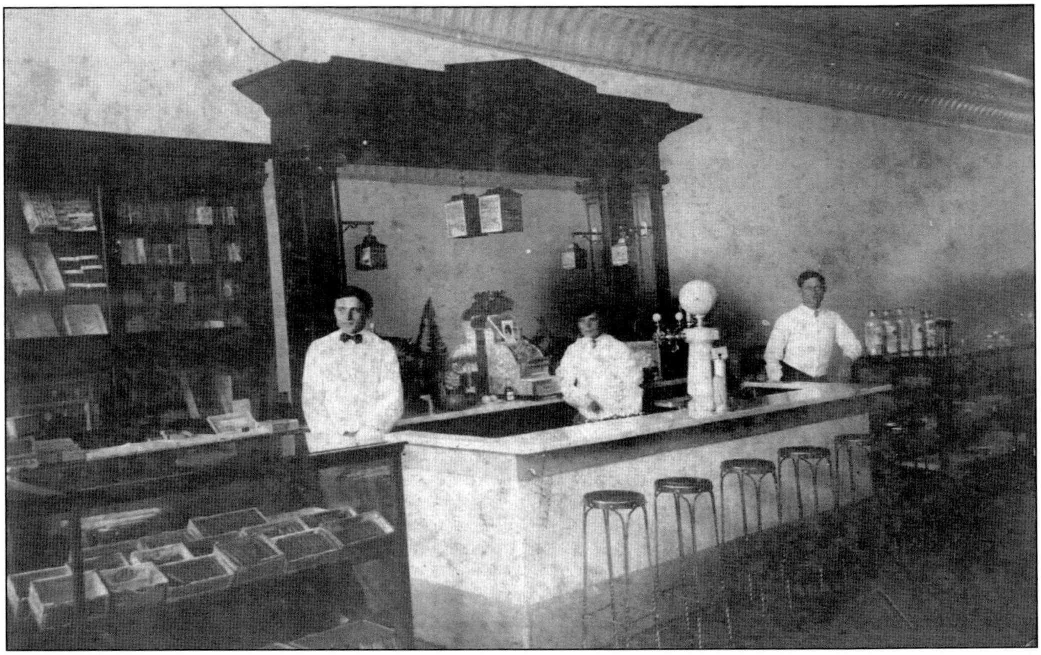

The McCree and Curry Place specialized in Cuban cigars. It was located in the brick Lamb Building, which also housed the Manatee County State Bank. Clerks are seen waiting to serve customers at the counters. The sophisticated décor was rare for shops in Palmetto, but workers attracted prominent businessmen visiting the bank. Note the bow tie worn by one of the workers to complete his uniform. (PHC.)

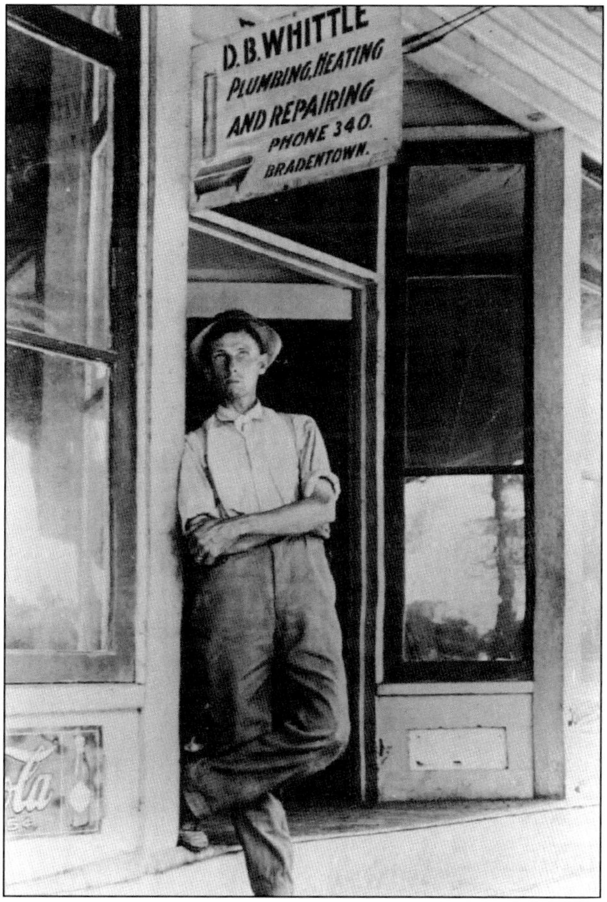

Palmetto resident D.B. Whittle stands outside his plumbing shop on Manatee Avenue. A year after this 1912 photograph was taken, he moved his business to Palmetto. During World War I, an irrigation pipe shortage left many area farmers in danger of losing their crops. Whittle invented a substitute. He constructed water pipes from eight-inch logs by cutting a hole through the log. The cores were removed and sold as fence posts. (PHC.)

The Oaks Hotel was built in 1909. The owners ran Palmetto's first sewer line from the hotel to the river, starting down an alley just west of Main Street. The sewer was available to any buildings on the west side of Main Street. For the privilege of tapping into the sewer, the city council gave the hotel two years of free water. The hotel advertised "old fashioned southern hospitality" to visitors. (Manatee Historical Reference Library.)

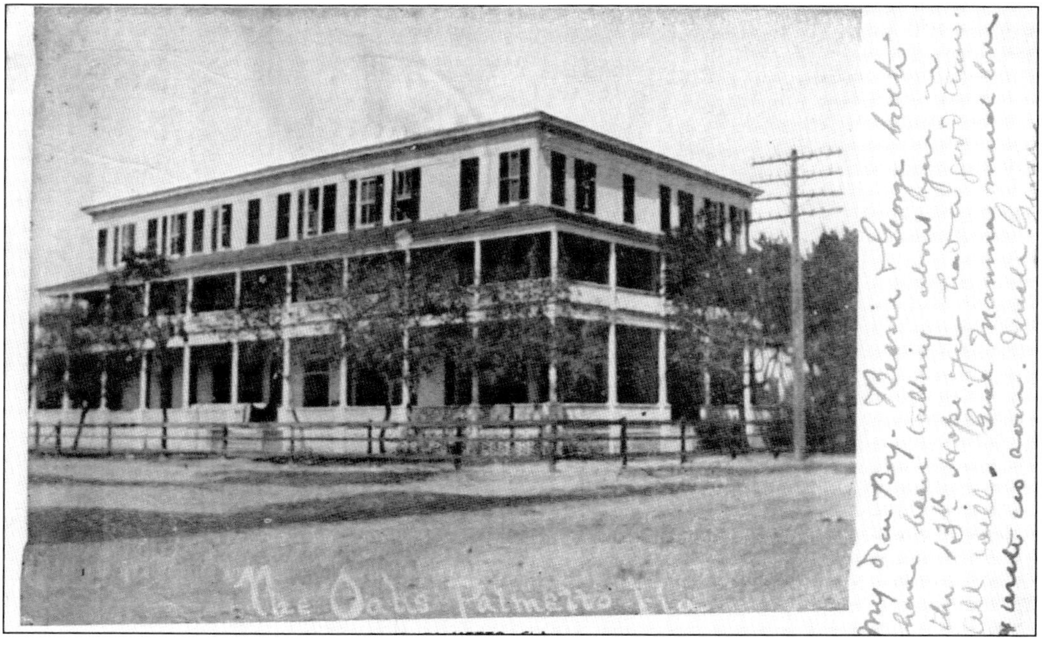

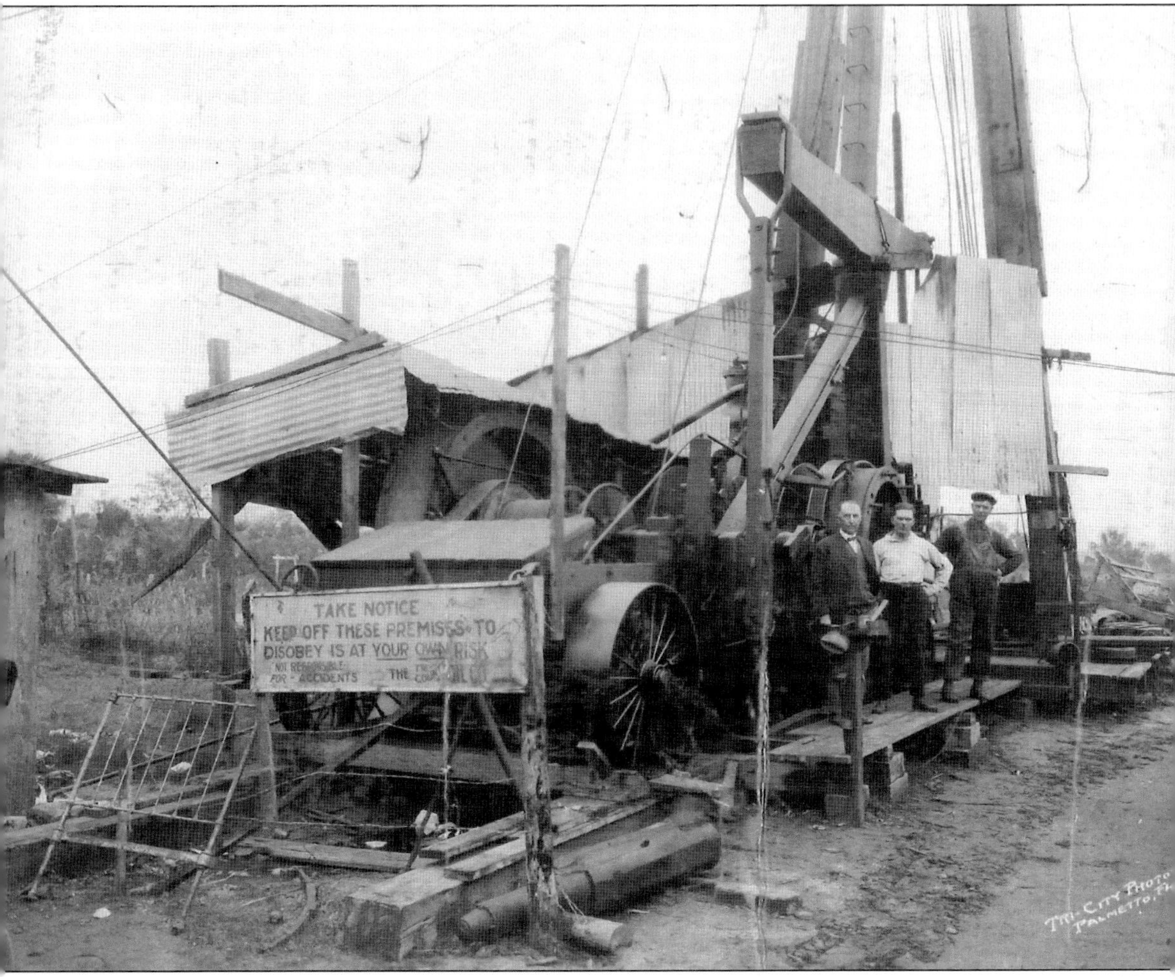

A year after the Tri-County Oil Company incorporated, an oil rig weighing 80,000 pounds arrived by train on February 14, 1924, accompanied by fanfare and the hope it would strike black gold. It was set up on W.E. Robinson's farm on Snead Island Road in the vicinity of the present-day Palmetto Police Department. On February 28, several hundred spectators gathered at the farm to observe the first drilling attempt. Following an optimistic speech by Capt. B.F. Alley—an oilman with an excellent reputation for discoveries—the 16-inch well brought up nothing but sulfur water. Three oil wells were attempted in Palmetto from 1924 to 1928. On the first attempt, the drill broke off at 1,300 to 1,400 feet. On the second attempt, the drill reached 2,600 feet, and traces of oil were discovered. However, on the third attempt, the drilling cable broke at 700 feet, and the expensive rig tools fell to the bottom of the well. The project was then abandoned, and the rig was used as scrap. (PHC.)

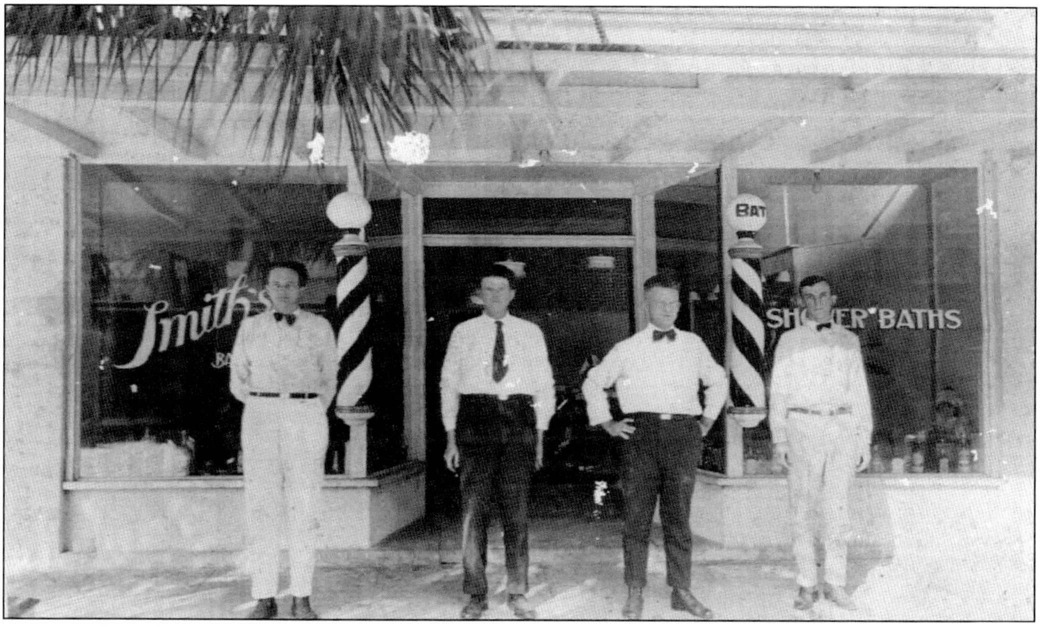

Four unidentified men stand outside of Smith's Barber Shop. English Smith and his wife, Addie, opened Smith's Department Store on Lemon Avenue (now Eighth Avenue) in 1924. They sold work cloths, patterns, and sewing machines. The barbershop was in the same building. In 1936, English opened a grocery next door for his son Willard "Cooter" Smith to manage. Smith's was open until 1999, when Walgreens bought the property. (PHC.)

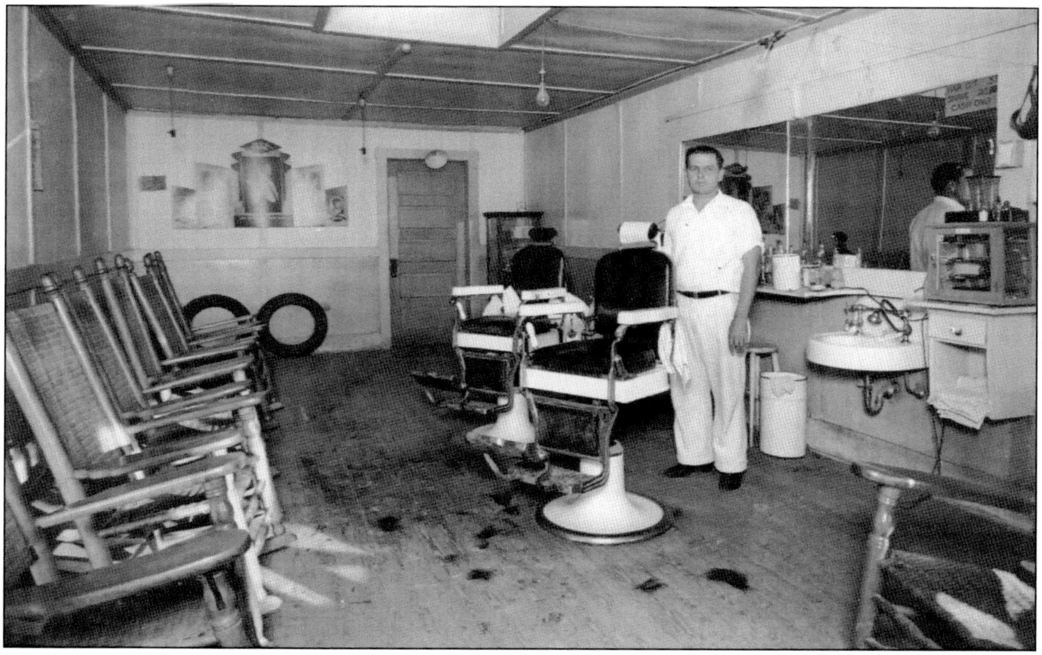

"Lowie the Barber" awaits his next customer at Ellis Barber Shop. Apparently, whoever sat in his chair right before this 1936 photograph was taken had a healthy head of hair. A row of rocking chairs offered customers a relaxing place to wait their turn, since there are only two antique barber chairs available. (PHC.)

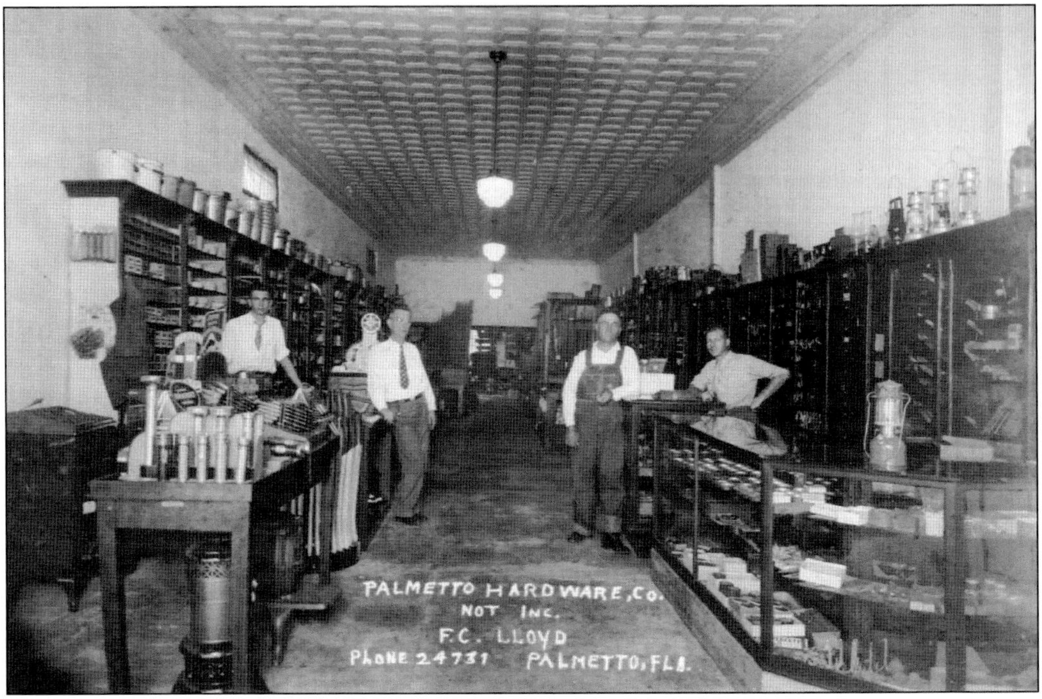

Charles Mathis and Foster Lloyd are recognizable in this 1918 interior photograph of Palmetto Hardware Company. The store was in the building that spans what is now known as 425–429 Tenth Avenue West, or Main Street. After housing the hardware store, a plumbing company occupied the building. (PHC.)

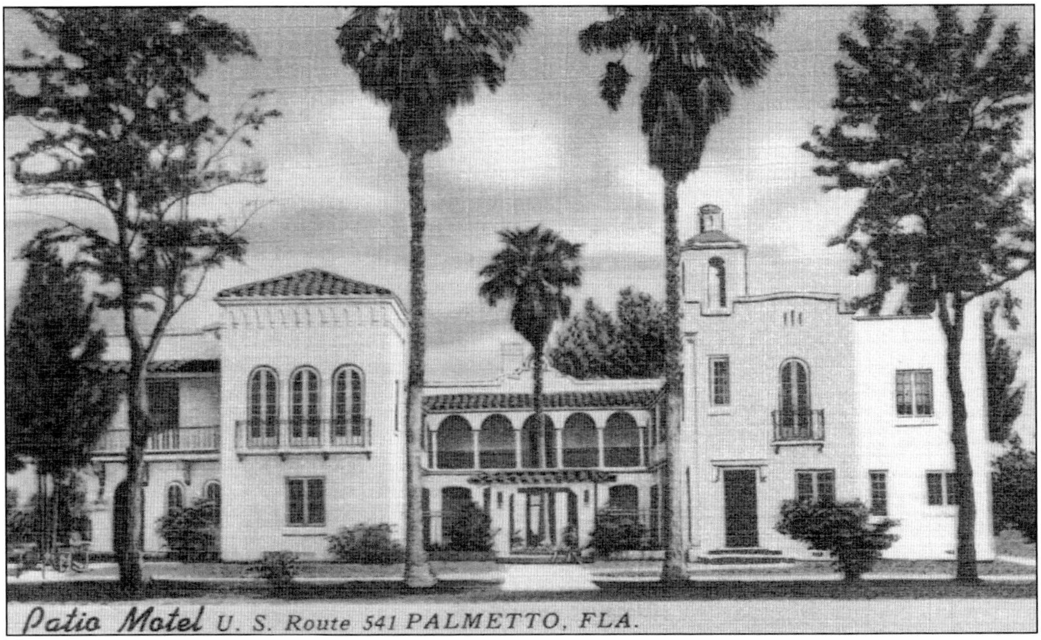

A postcard by Harold R. Smith features the luxurious Patio Hotel on US 41, just 1.5 miles north of Palmetto. The Spanish-style building had 17 units for rent. A. Zaebarios was the manager of the hotel. (PHC.)

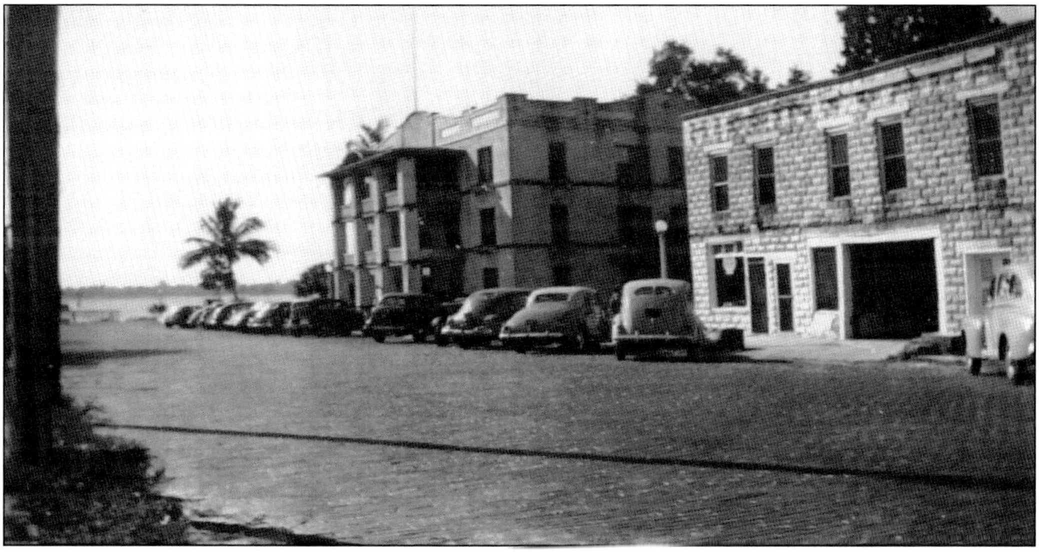

On the right is the Stoltz apartment building and Higgins Garage and tire shop on Main Street. The Stoltz Building is now known as the Crusader Building after it housed a Canadian sailing team by the same name that trained for the America's Cup in the area. (PHC.)

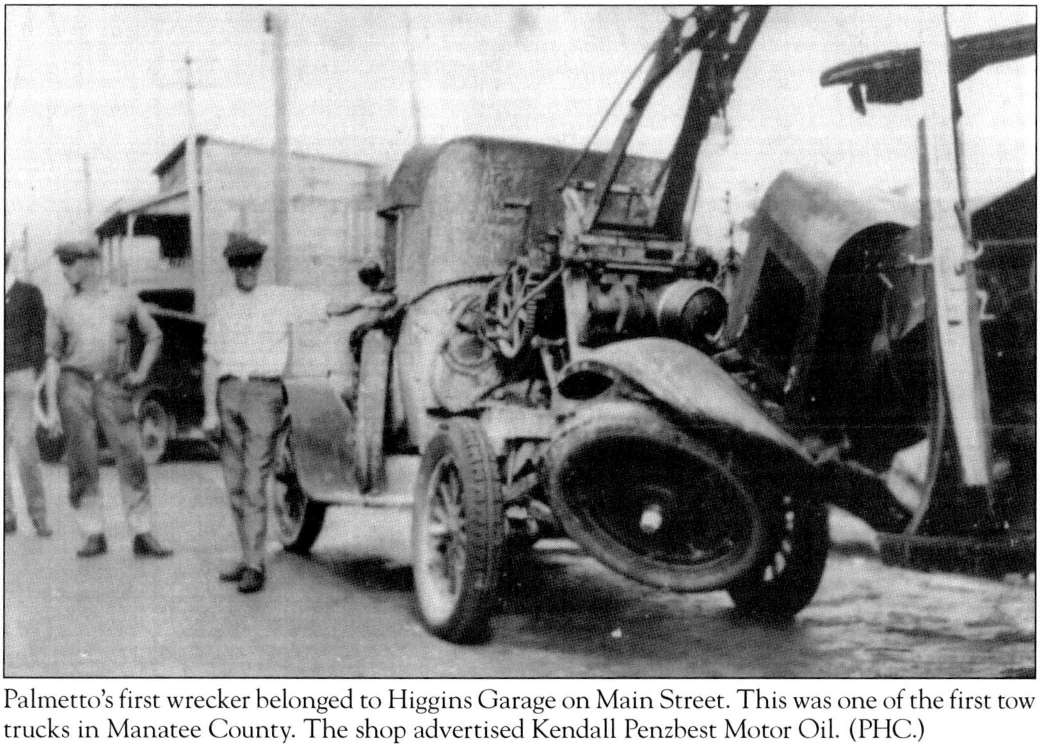

Palmetto's first wrecker belonged to Higgins Garage on Main Street. This was one of the first tow trucks in Manatee County. The shop advertised Kendall Penzbest Motor Oil. (PHC.)

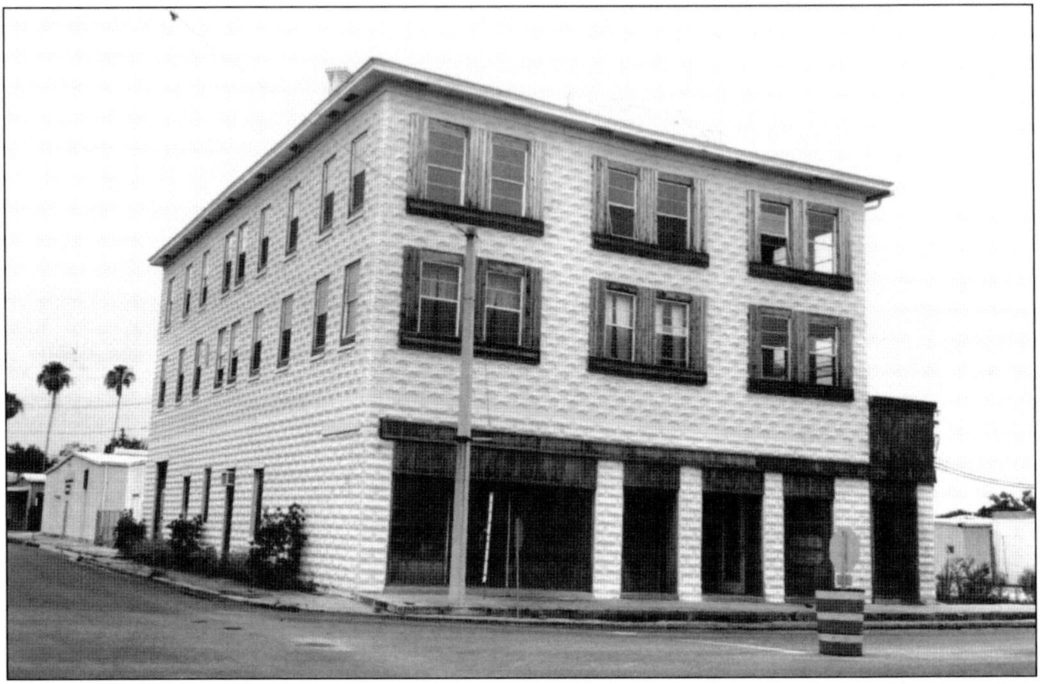

The Washington Restaurant was located at 815 Eighth Avenue and was owned by Mike Mavroleo. The restaurant specialized in western steaks, pork chops, chicken, and yellow rice. For dessert, it served homemade pies. (PHC.)

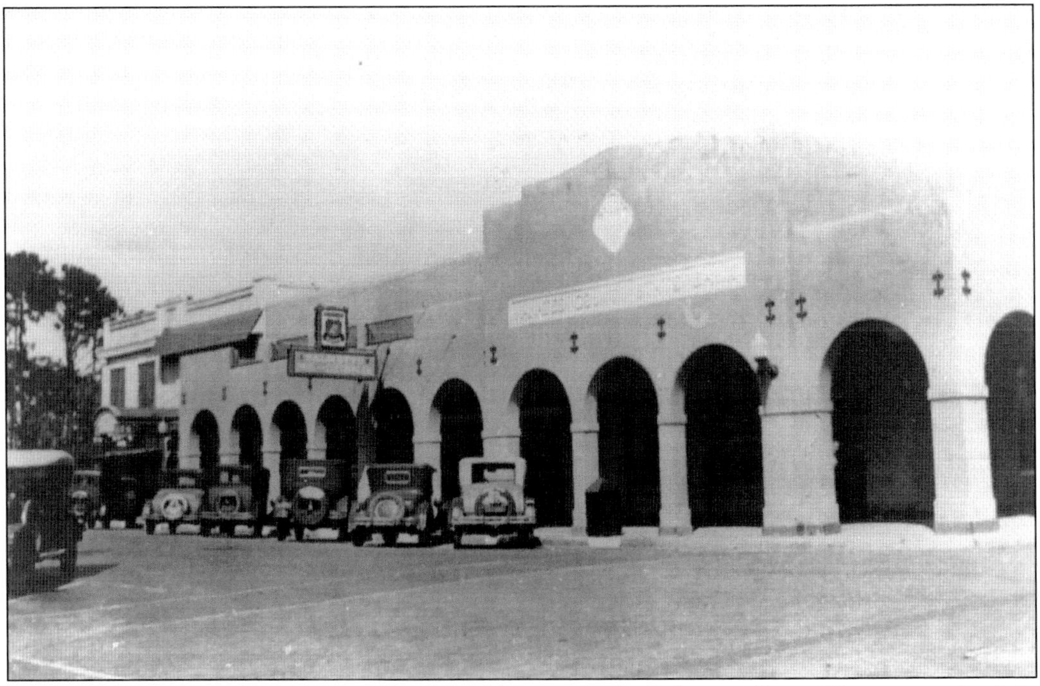

To the right is the Palmetto State Bank building with the Palmetto State Theater on the left. Thomas F. Taylor, or "Taylor the Tailor," owned the movie theater and vowed no child would ever pay more than a nickel to see a movie in his theater. (PHC.)

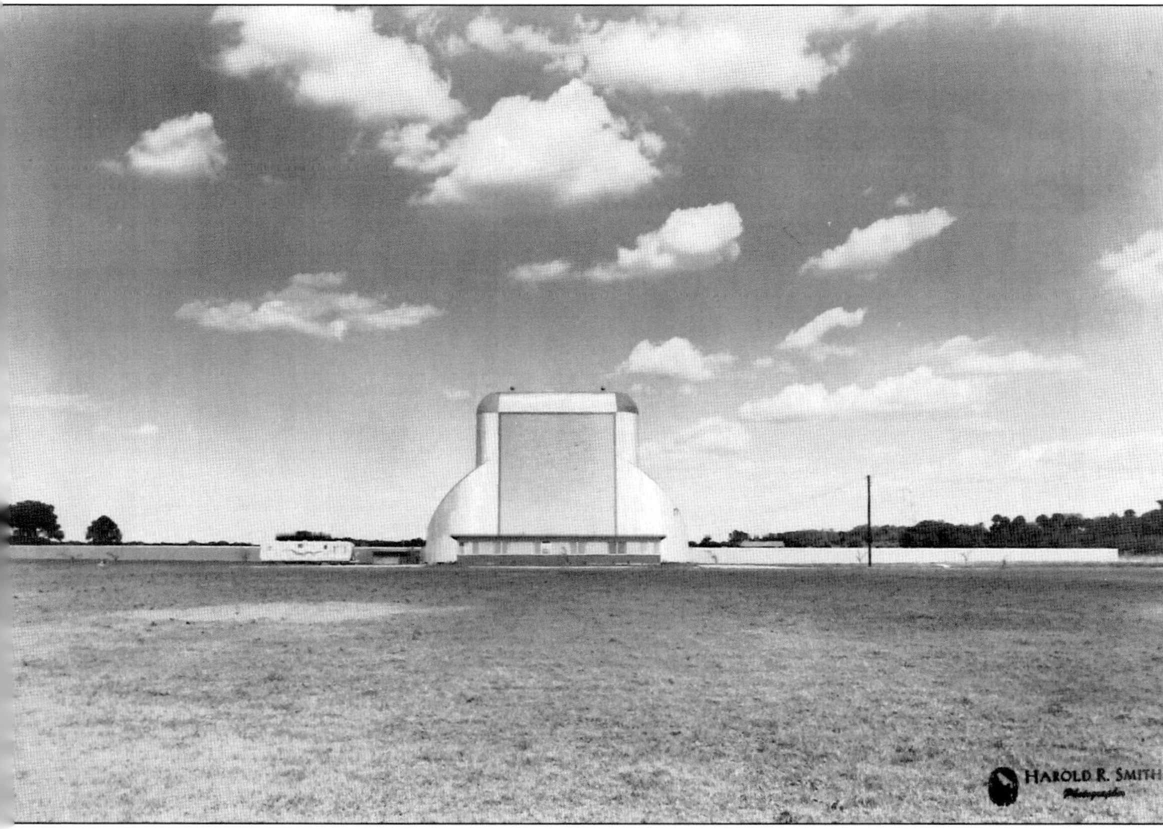

Harold R. Smith took this photograph in 1951. The Palmetto Drive-In was constructed in 1951 and owned by Roy and Kathleen Jones. Latino customers were allowed to watch movies here, something that did not sit well with the local Ku Klux Klan. A cross was burned at the theater's entrance on US 41. Roy claimed the Klan also tried to scare him off by putting poisonous snakes in his Cadillac. On Halloween night in 1958, a tornado destroyed the theater, as well as the outer buildings where the Jones lived; all were rebuilt. On Sundays, a drive-in church was held with a non-denominational service and served free coffee and donuts to everyone attending. The preacher sat on a platform, and the congregation could listen in from their cars. Although the drive-in was demolished in 1986, Kathleen opened a new drive-in church just north of the original theater site on US 41 that still operates today. (Courtesy the Jones family.)

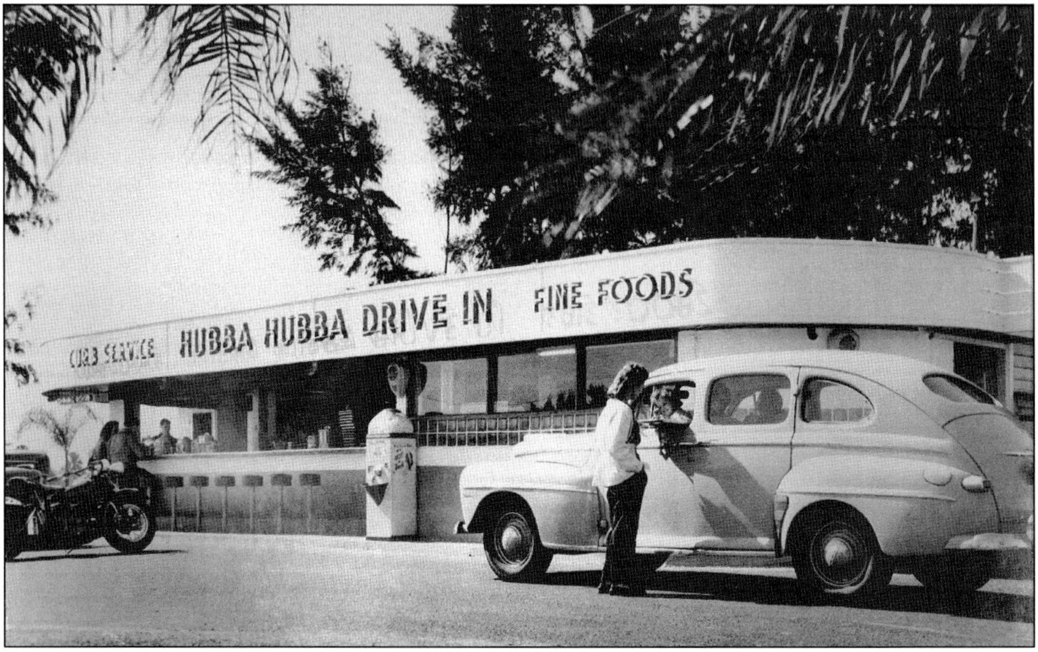

The Hubba Hubba Drive-In opened in the 1930s at the corner of Seventeenth Street and Eighth Avenue. On the roof stood a hula girl sign that swayed in the breeze and was considered quite scandalous among religious institutions. The hula girl became an unofficial Palmetto trademark—it was the first view visitors traveling on US 41 had of the city. The popular landmark was torn down in 1974. (PHC.)

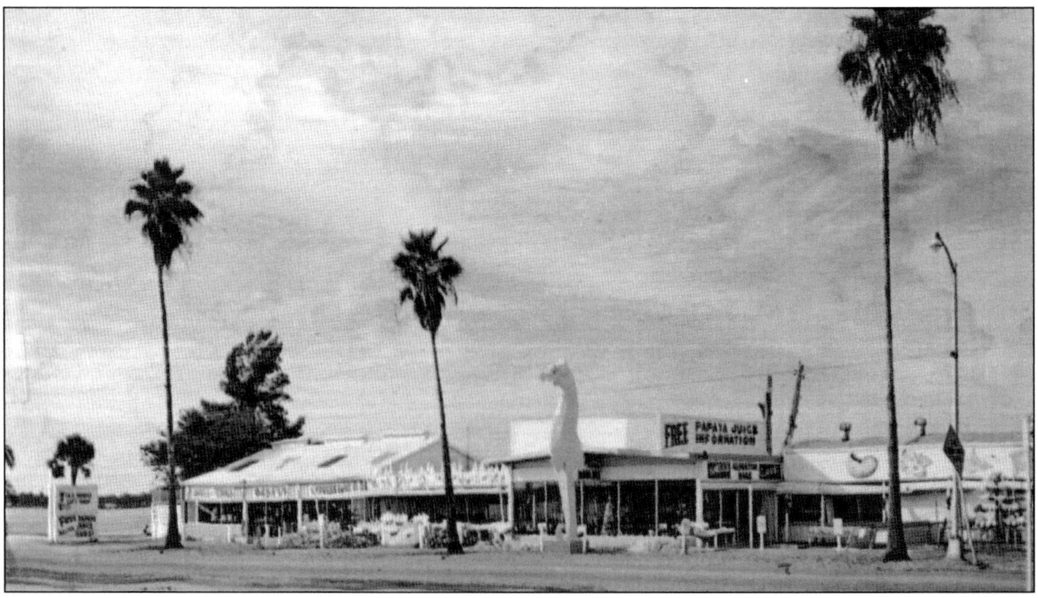

Ed and Evelyn Winter opened the Seahorse Gift Shop near the Green Bridge in 1953. Specializing in converting shells to gifts, the employees designed anything from earrings to table lamps. The Seahorse had a pine tree growing through it. Ed Winter's menagerie of "pets" on display, including alligators, rattlesnakes, monkeys, and a 50-year-old macaw named Scarlett, helped draw customers in. Ed astounded tourists by milking rattlesnakes in the parking lot. (PHC.)

Five

A Struggle for Survival

As soon as they set foot on their sultry, scrub homesteads, Palmetto's pioneers began to face a series of struggles: wars, disease, and natural disasters among them.

Congress passed the 1842 Homestead Act (formally "An Act to Provide for the Armed Occupation and Settlement of the Unsettled Part of the Peninsula of East Florida") to lure settlers to Florida with a promise of 160 acres of free land. But a daunting condition applied: homesteaders had to remain on the land for five years before they owned it outright. A series of wars with the Seminoles had deterred settlement, so the government hoped the new arrivals would bring stability and ward off the Seminoles. As it happened, some homesteaders literally paid for a piece of Florida with their own blood. In 1855, word of renewed Seminole attacks swept across the water to weary settlers, and heads of households were obliged to form militias, leave their farms in the hands of their families, and go to battle.

Just two years after their return from fighting the Seminoles in the Everglades, the men of Palmetto again packed muskets in preparation for the Civil War. The Union blockade halted the profitable Cuban cattle trade, and federal troops destroyed local sugar plantations suspected of providing supplies to Southern troops. Near the war's end, the rebels recruited homesteaders to drive cattle north for a starving Confederate Army. Slaves who fled their masters were taken to Egmont Key, where they trained to serve in the Union Army.

In 1888, the yellow fever epidemic brought the first doctors to Palmetto. Until yellow fever hit, residents considered the climate therapeutic for a range of ailments, and a good number of them came to Florida for its healing benefits only to face deadly outbreaks of disease. The earliest known physician to serve Palmetto was Dr. John Crews Pelot of Braidentown, who would cross the river at Rye Bridge and ride into town on horseback when summoned. When residents began dying, a puzzled Dr. Pelot set up sawhorses and performed an outdoor autopsy in a field on the corner of Eighth Avenue and Seventh Street, but he could not determine the origins of the disease, later diagnosed as yellow fever. Dr. M.B. Harrison also came to town to minister to the sick and dying, and he became the first physician to open a practice in Palmetto.

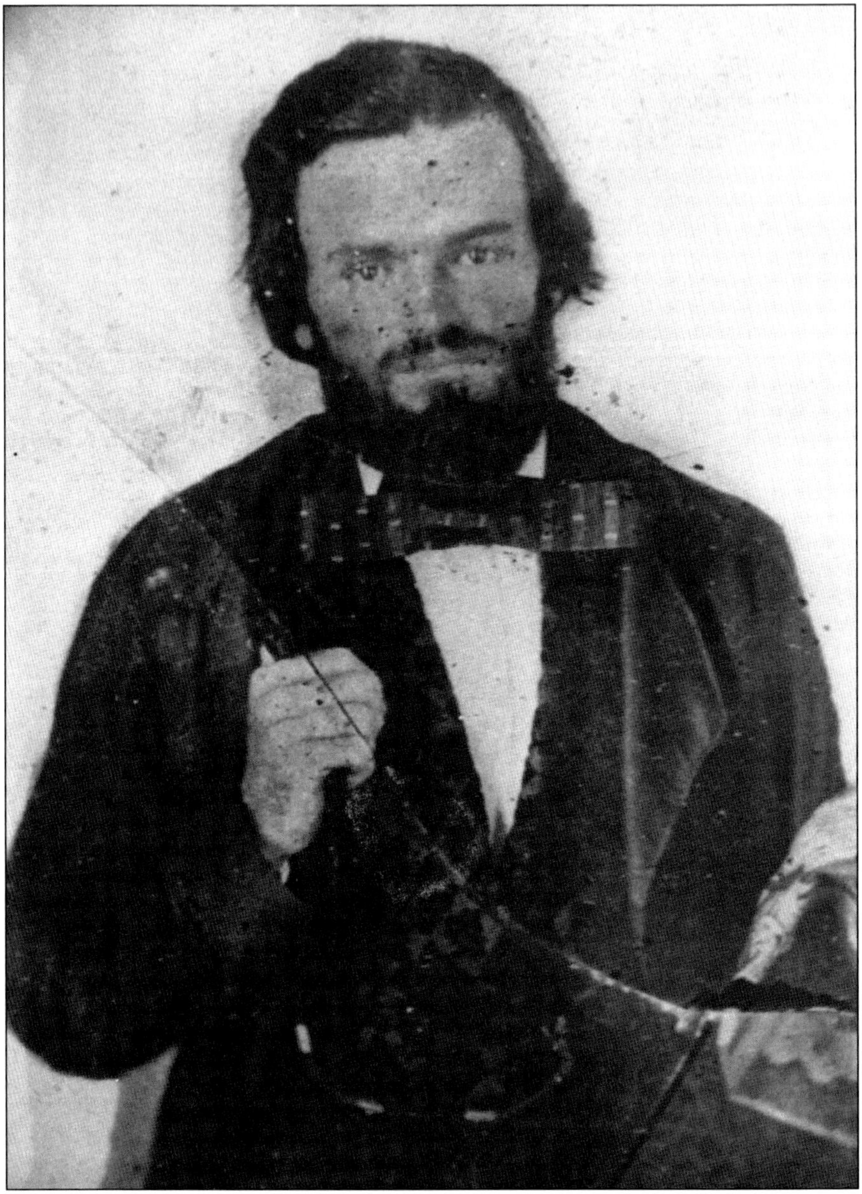

Under the direction of Capt. John T. Lesley (pictured) of Tampa, three companies were organized, armed, and equipped with rations to fight the hostile Seminoles during the third Seminole War (1855–1858). Lesley joined the Florida militia as a private but was quickly promoted to lieutenant. He handpicked 31 men from Manatee, Hillsborough, and Benton (now Hernando) Counties based on their familiarity with native customs. Among them, Palmetto residents Joe Atzeroth, Joseph Braden, John Craig, Daniel Hawkins, and Simon Snell were employed as boatmen to track wanted refugees through the swamps of the Everglades, where no horseman could go. A female Seminole guide named Polly aided them in their pursuit. Lesley's outfit shipped 139 captured Seminoles, including the infamous Seminole leader Billy Bowlegs I, to Egmont Key. From there, the Seminoles were forced to embark on a solemn sail to New Orleans and join other Native Americans on the Trail of Tears—the harsh, forced march for their relocation in Indian Territory in present-day Oklahoma. (SLAF.)

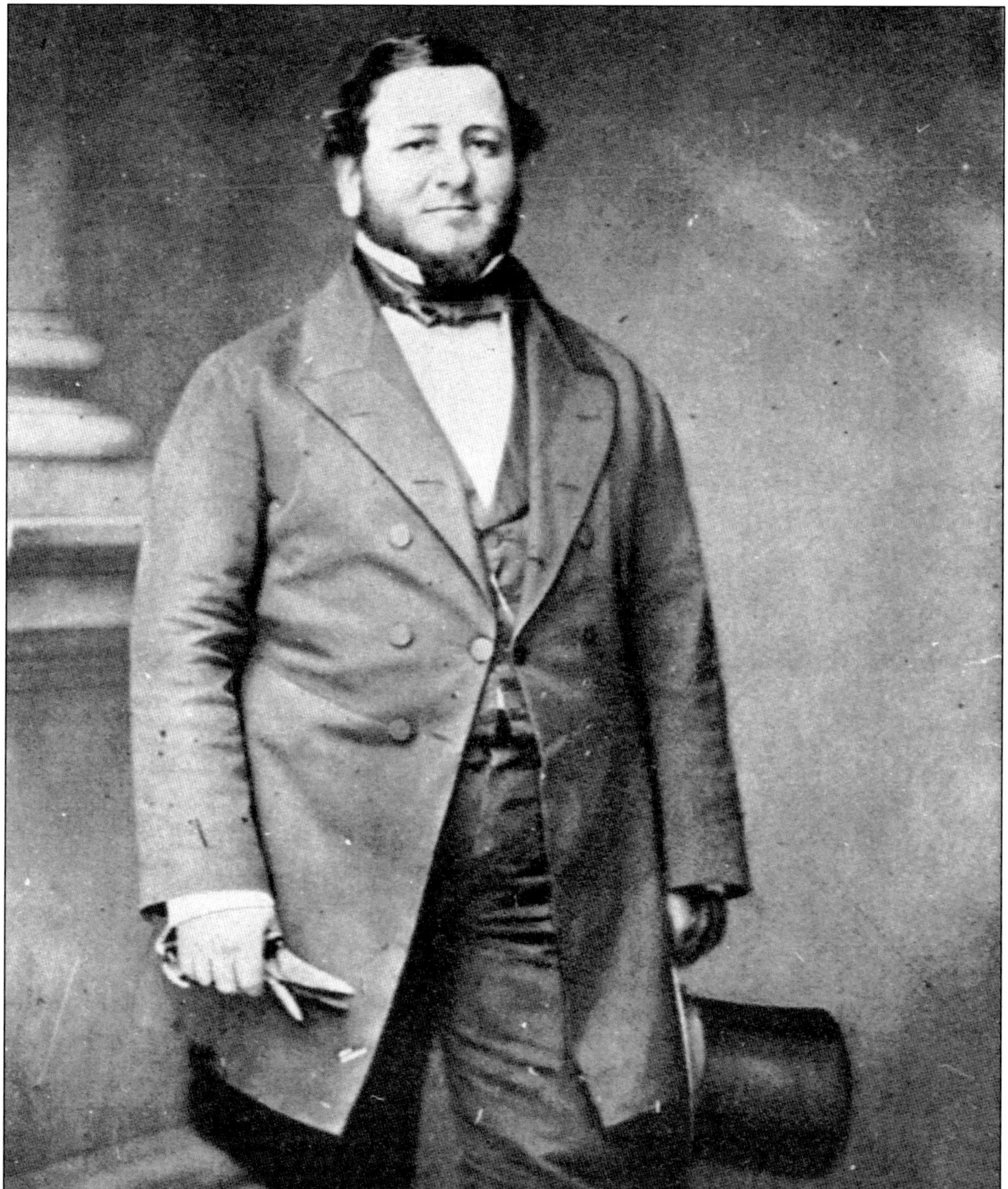

At the conclusion of the Civil War, Judah P. Benjamin, secretary of state for the Confederacy, fled to Florida disguised as a Frenchman named Monsieur Banfal. He assumed other identities, including pretending to be a farmer looking for land to settle. The Union posted a $50,000 reward for his capture. Benjamin made his way to the Gamble plantation, then a sanctuary for wanted men, and met Capt. Archibald McNeil, a Scottish navigator, who himself was wanted for blockade running. Benjamin occupied a room overlooking the Manatee River, from which he kept a watchful eye for Union gunboats searching the shores for fugitives. After a close encounter with a search party, Benjamin hid in a meat cart and was smuggled to Sarasota, where Capt. Fredrick Tresca waited to aid in Benjamin's escape. They departed for Cuba on Tresca's yawl, the *Blonde*. Benjamin later traveled to England, where he became a barrister. (MCHS.)

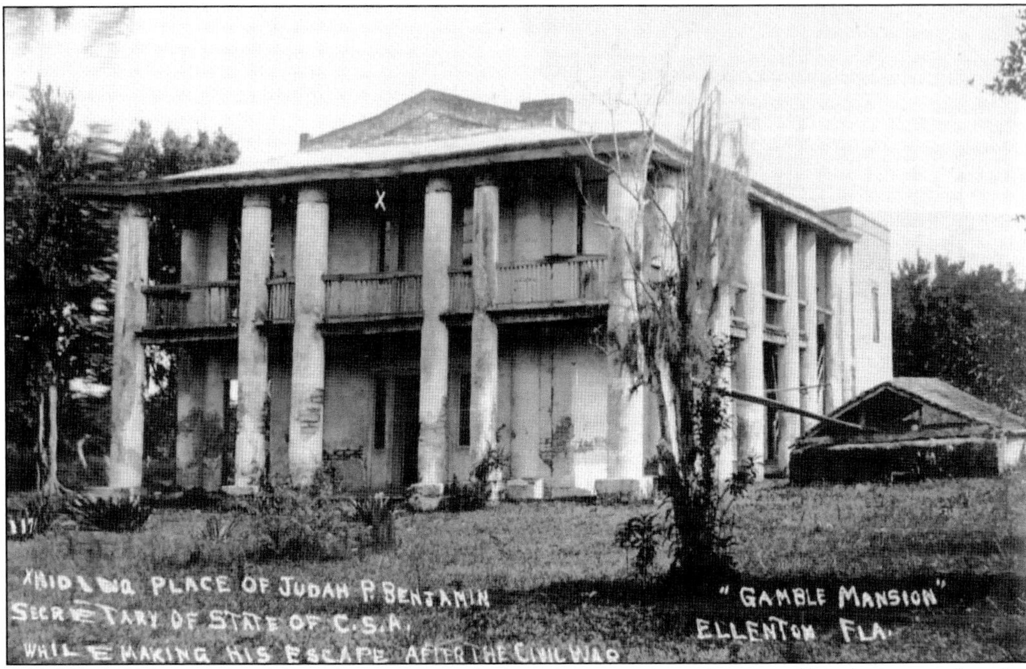

The Gamble plantation barely survived the Civil War. Union soldiers spared the mansion during a raid in 1864, but the sugar mill and saltworks near the premises were destroyed. The *x* marks the room occupied by Judah P. Benjamin. The estate was purchased by Maj. George Patten in 1873 and was occupied for 20 years. It was deeded to the state in 1925 and became a state park. (PHC.)

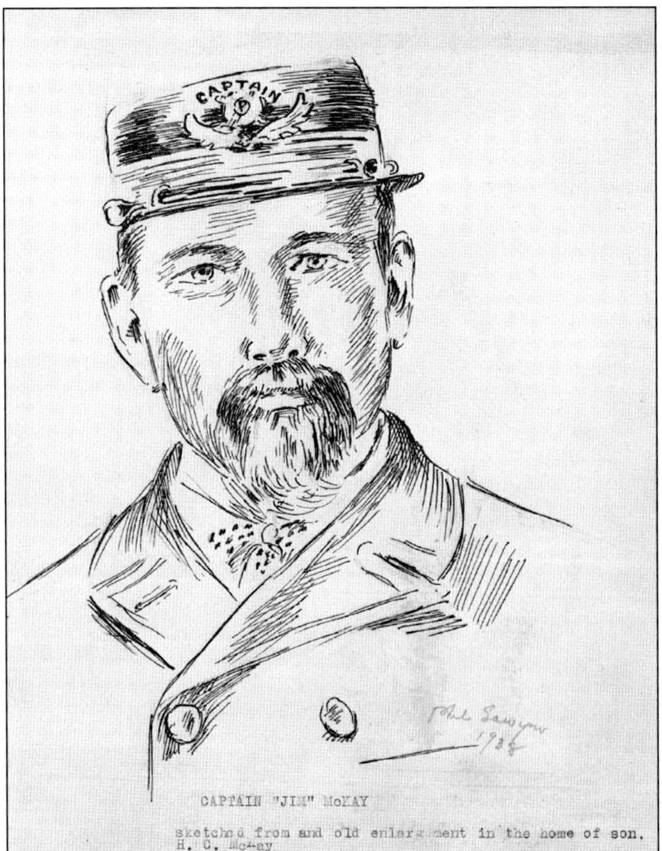

Capt. James McKay, the first commercial cattle shipper, became the commissionaire agent for the rebels assigned to get cattle and food to the Confederate Army. His steamer *Fargo* was captured on June 7, 1861. McKay was thrown in prison at Fort Jefferson. After five months, he escaped and continued to evade Union troops, operating a saltworks in Ellenton that was destroyed during a raid on August 4, 1864. (PHC.)

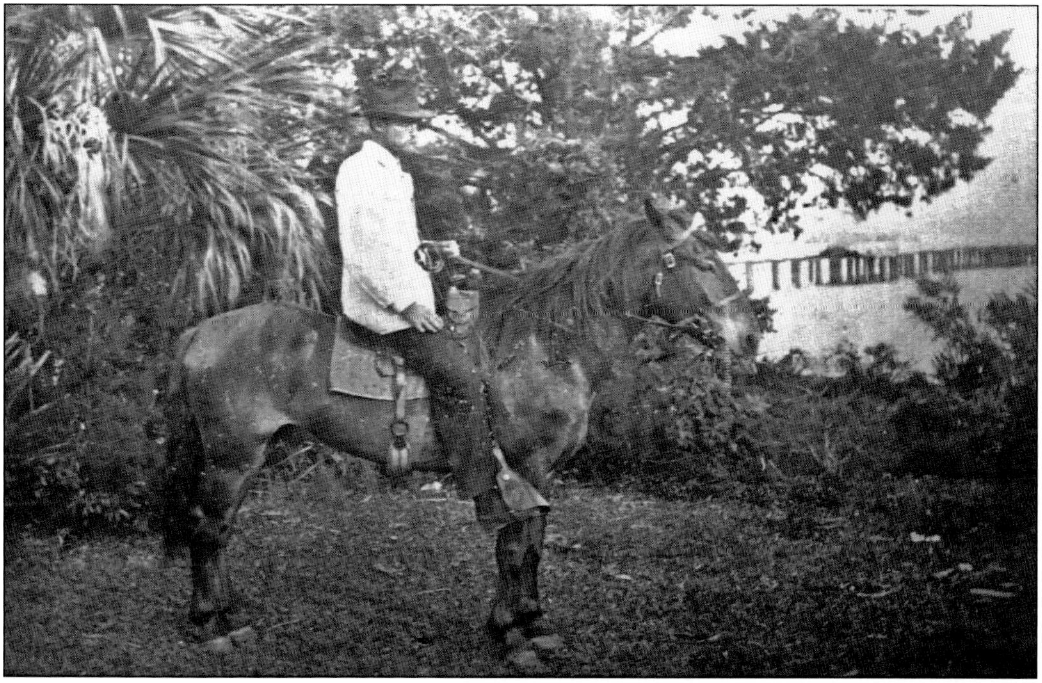

Dr. John Oliver Brown came to Palmetto in 1893 at the age of 24 to practice medicine. He was particularly interested in typhoid and malaria fevers. He ordered open-water wells filled and all rainwater cisterns covered, because he believed it would reduce the mosquito population. He served the community for 57 years, delivering 976 babies and traveling as many as 30 miles on horseback to tend to his patients. (PHC.)

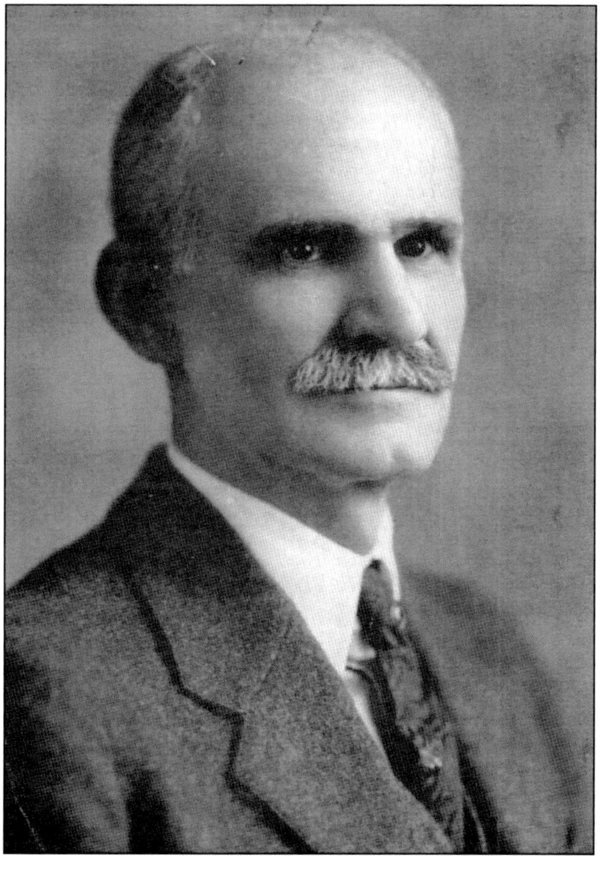

Dr. John S. Helms arrived in Palmetto in 1896. He noticed immediately that many of the area's children featured sallow complexions and behaved lethargically. He discovered they were suffering from hookworm, an often-fatal ailment that plagued many Southern states. A simple step of better sanitation eradicated the parasite. Helms later opened a drugstore at Ninth Avenue and Fourth Street. (Courtesy Ken Burton.)

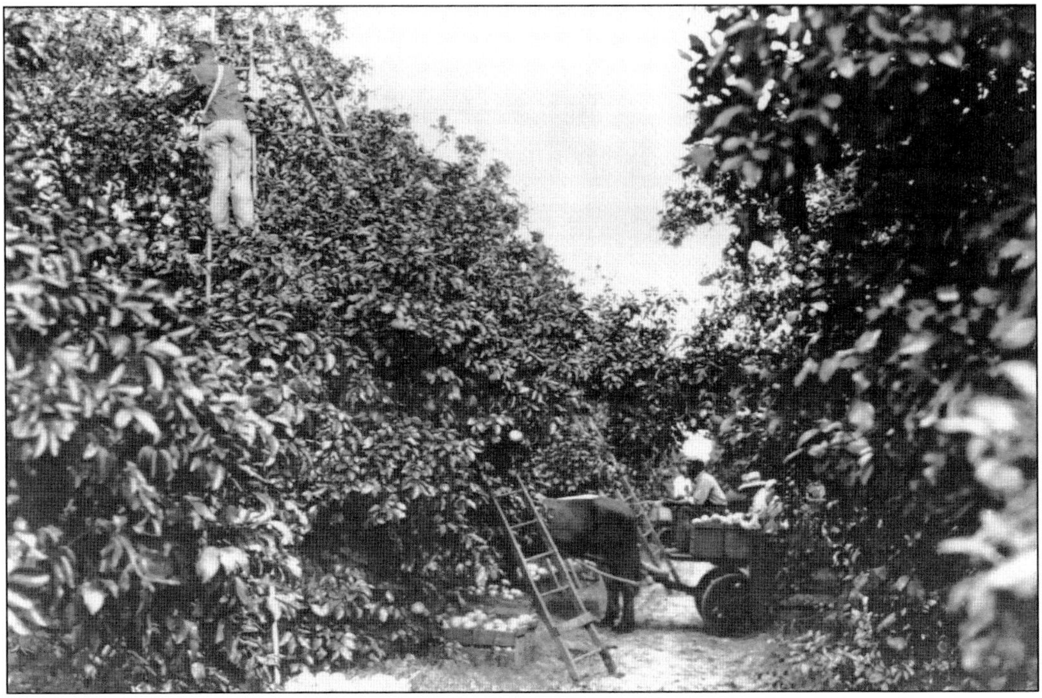

In 1896, the local KKK doled out a weeklong reign of terror directed at C.L. Harvey, manager of the Manatee Lemon Company, and his African American workers, who were told to leave town. An altercation erupted into gunfire when 15 men attempted to seize and punish the 14-year-old son of Jack Trice. Jack resisted and shot four members of the mob, killing town marshal James Hughes before fleeing. Trice was regarded as a hero in the African American community. (MCHS.)

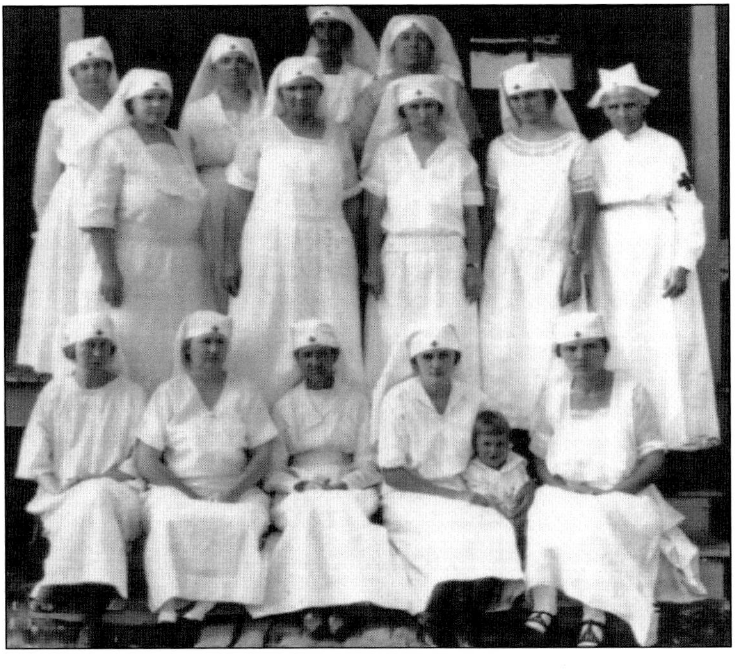

Prior to World War I, the Red Cross introduced its first-aid, water-safety, and public-health nursing programs. With the outbreak of war, the organization experienced phenomenal growth directing its focus to military aid. The number of local chapters jumped from 107 in 1914 to 3,864 in 1918, and membership grew from 17,000 to more than 31 million. This photograph of Palmetto's Red Cross nursing school was taken in 1918. (PHC.)

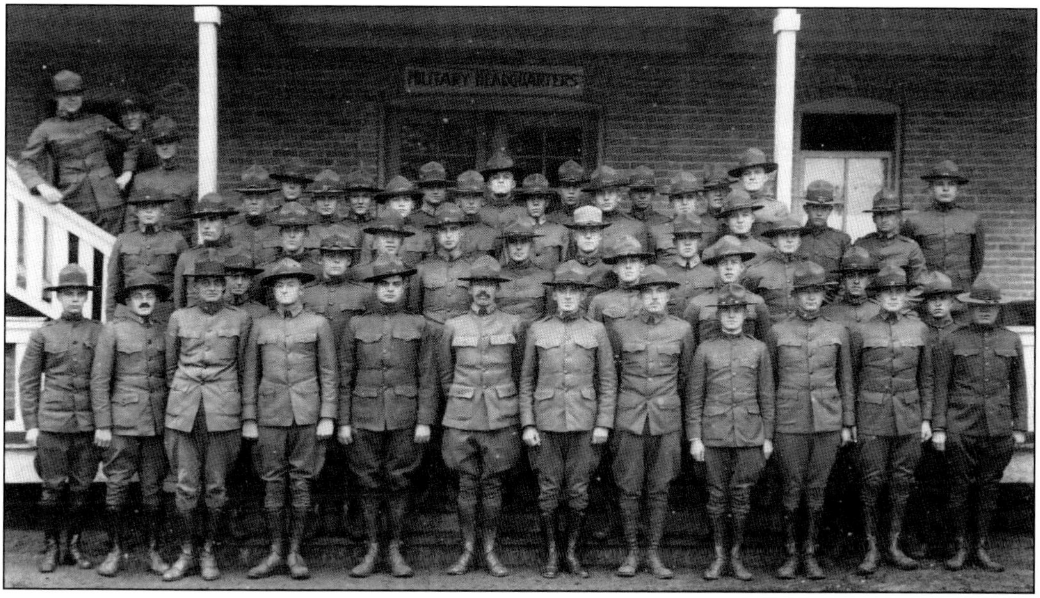

James A. Howze II, son of Capt. James A. and Frankie Howze, is the farthest to the right in the front row. On April 6, 1917, the United States declared war on Germany and its allies. Howze enlisted in the US Army on July 1. He was assigned to Camp Greenleaf in Georgia in 1918. Howze was one of the 143 Palmetto residents to serve in the Great War. (PHC.)

2

Johnson, James (Surname) (Christian name) 4,186,250 (Army serial number) ~~*White~~ *Colored.

Residence: xx (Street and house number) Palmetto, Manatee Co. (Town or city) (County) FLORIDA (State)

~~*Enlisted *R. A. *N. G. *E. R. C.~~ *Inducted at Bradentown Fla on Aug 3, 1918

Place of birth: Dumellon Fla Age or date of birth: June 25/94

Organizations served in, with dates of assignments and transfers: 62 Co 13 Tng Bn 151 Dep Brig to Aug 26/18; Co L 807 Pion Inf to disch

Grades, with date of appointment: Corp-; Sgt Dec 4/19

Engagements: xx

Wounds or other injuries received in action: None.

Served overseas from † Sept 4/18 to † Apr 19/19, from † to †

Honorably discharged: Sept 4, 1920 (Date) S.C.D. (Cause)

In view of occupation he was, on date of discharge, reported 50 per cent disabled.

Remarks:

Sgt. James Johnson was one of 51 African Americans in Palmetto to serve in World War I. Deployed overseas in the 92nd or 93rd Division, the only black combat outfits, Johnson belonged to a labor unit. Its responsibilities included digging trenches, removing unexploded shells, clearing disabled equipment, and burying soldiers killed in action. Despite the hard work they provided, African American stevedores received the worst treatment of all black troops serving in World War I. (SLAF.)

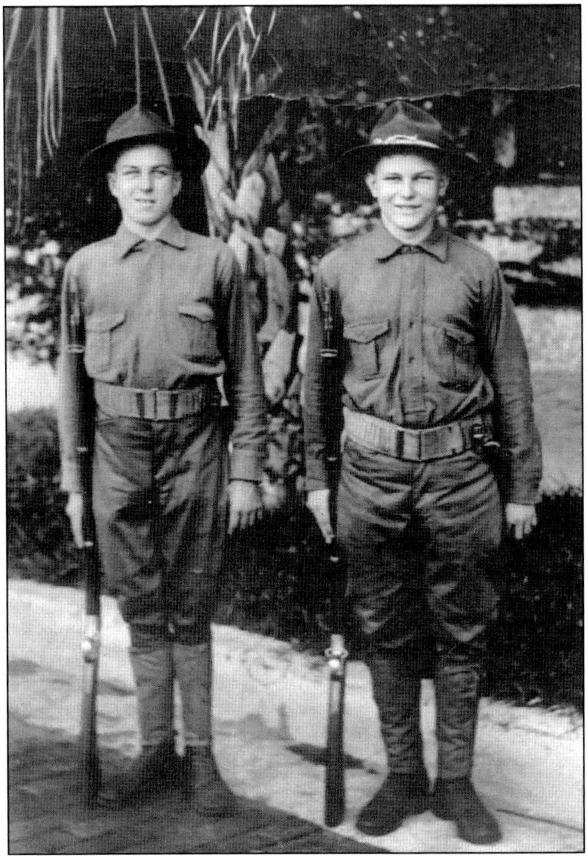

Bill Snyder and Stuart Howze were members of the Home Guard in 1918. That year, thousands of US soldiers were deployed to Europe to combat the German Empire. While all of Florida's guardsmen were in federal service, the state established a system of county guards to perform active-duty service on the home front. On January 1, 1919, Florida's entire official military force was comprised of 159 men in guard armories around the state. (PHC.)

Sgt. Edward L. Freyermuth of Company C, First Field Signal Battalion, Second Division, received the French Croix de Guerre with a silver star on February 14, 1919. The medal is bestowed on foreign military forces allied to France and Belgium for acts of heroism involving combat with enemy forces. According to military documents, Freyermuth voluntarily went to the front line to assist with counter attacks at a critical moment. (SLAF.)

1½

Freyermuth (Surname) **Edward L** (Christian name) **198,288** (Army serial number) **White** (Race: White or colored)

Residence: -- (Street and house number) Palmetto (Town or city) FLORIDA (County) (State)

*Enlisted in R A at Ft McPherson, GA., May 24, 1917.

†Born in Ocala, Florida Age 22 4/12 yrs.

Organizations: Dep Co 1, Sig C. to June 26/17; Co C 1 F Sig Bn to Feb 12/19; Co B 1 F Sig Bn to July 13/19; Co B 322 F Sig Bn to Aug 4/19; Hoboken Cas Co #1011 to disch.

Grades: Pvt 1/c Oct 24/17; Corp Dec 17/17; Sgt Oct 16/18

Engagements: Troyon Sector; Toulon-Tryon Sectors; Aisne Defensive; Chateau-Thierry Sector; Aisne-Marne Offensive; Marbache Sector; St Mihiel Offensive; Meuse-Argonne Offensive; Meuse-Argonne (Champagne) Offensive.

Wounds or other injuries received in action: None.

‡Served overseas: Dec 14/17 - Aug 3/19

§Hon. disch. Aug. 14/19 on demobilization

Was reported 0 per cent disabled on date of discharge, in view of occupation.

Remarks: Awarded Fr.C.de G.with-silver star.

Form No. 724-1½, A. G. O. March 12, 1920. 3—7688

*Insert "R. A.", "N. G.", "E. R. C.", "N. A.", as case may be, followed by place and date of enlistment. †Give place of birth and date of birth, or age at enlistment. ‡Give dates of departure from and arrival in the United States. §Give date.

To halt the illegal production of whiskey during Prohibition, Mayor L.P. Thomas and the Palmetto City Council enlisted the aid of James Edmund Bowdoin. Bowdoin, a US Treasury Department agent, earned a reputation as a crime fighter for ridding other communities of stills and moonshiners. His gun collection and the pearl-handled pistols he wore at each hip earned him the nickname of "Pistol Pete." He never let jurisdictional lines or legal formalities stop him from making arrests. Moonshiners had clever methods of concealment, such as building hog pens over a still to mask the odor of fermentation. They seldom fooled Bowdoin. On one particularly successful afternoon, Bowdoin found and destroyed 11 stills. Although his stay in Manatee County was brief (1922–1924), Bowdoin's exploits became legendary; he would often load the confiscated still on his car and drive into town for all to see. (PHC.)

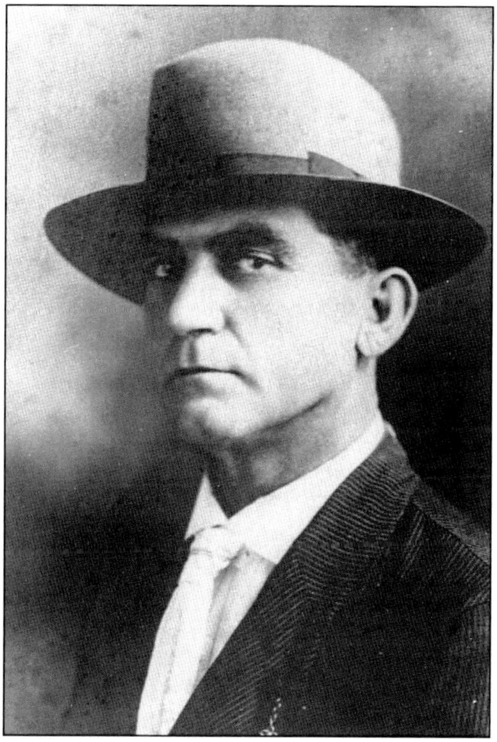

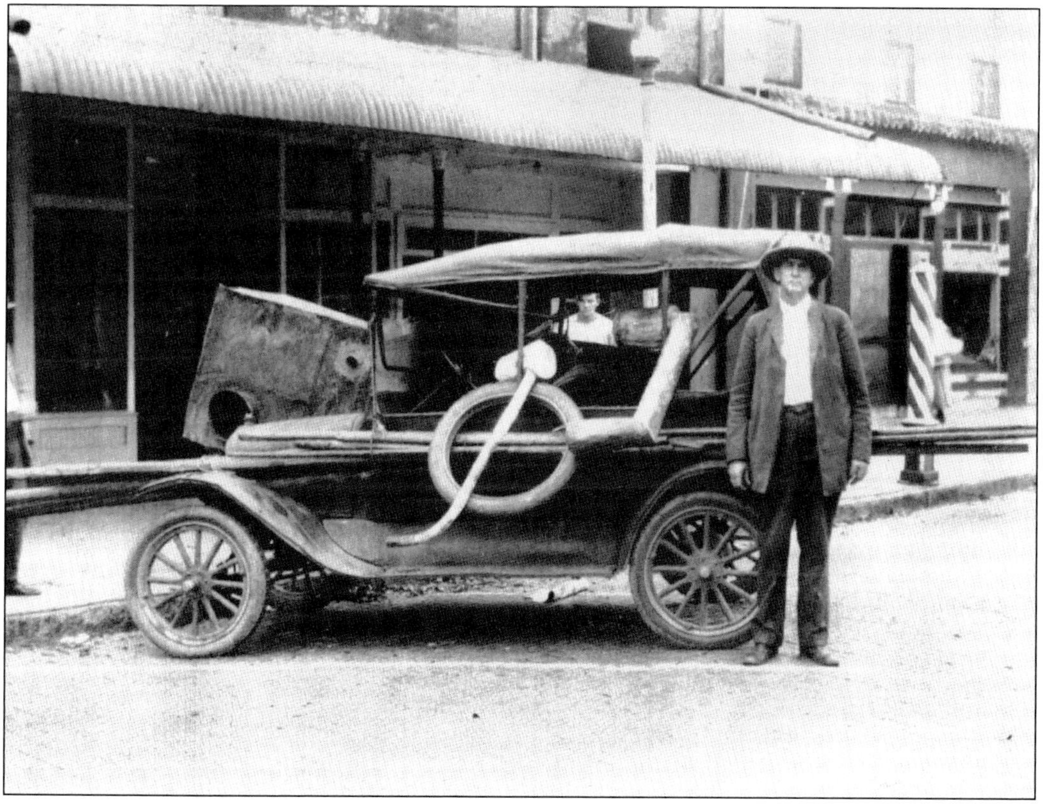

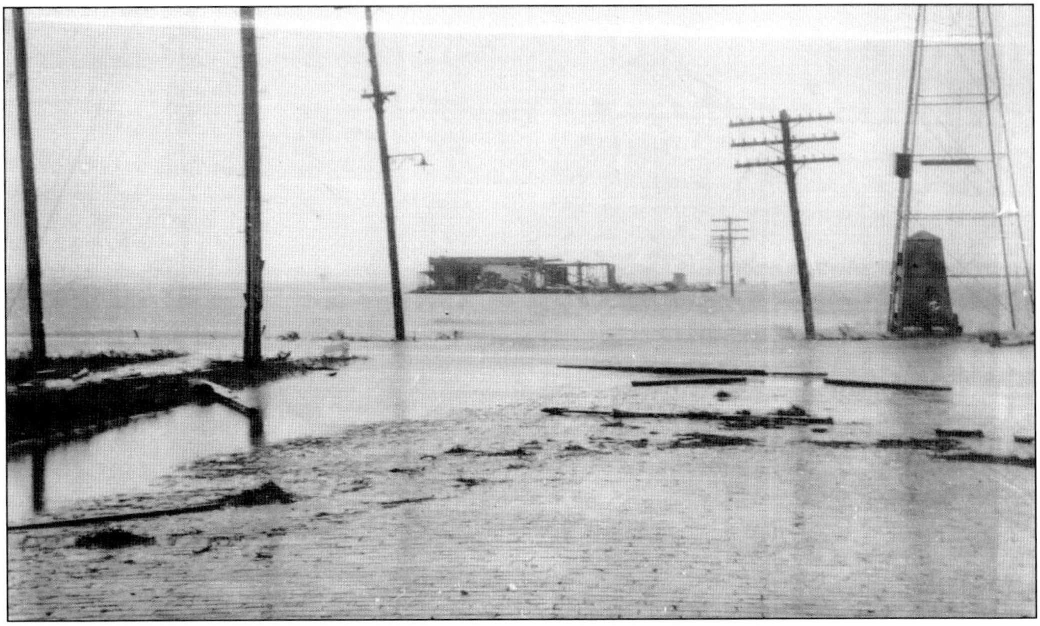

Hurricanes have been a constant threat to the thriving town of Palmetto. The first documented hurricane struck in 1846, sweeping down the Manatee River from a northeasterly direction. In the morning, witnesses said the retreating storm nearly sucked the river dry. In one account, a deer was seen crossing the river without having to swim. In 1921, a hurricane caused mass destruction, destroying everything in its path and leaving behind the remnants of City Dock. Below, a destroyed Palmetto building illustrates the storm's power. (PHC.)

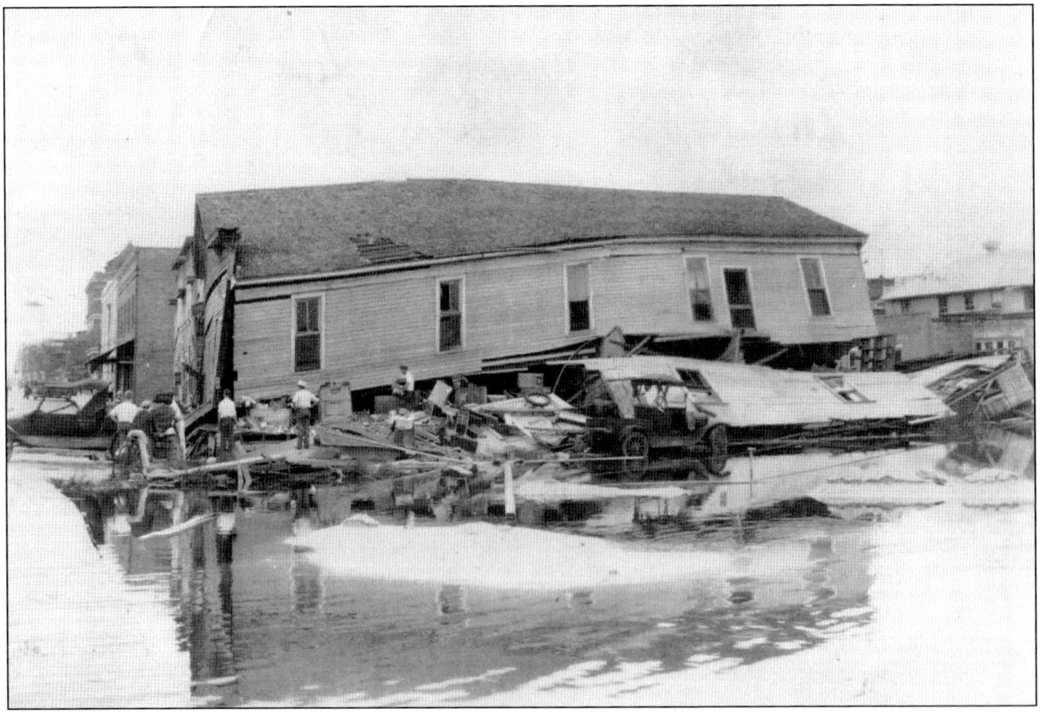

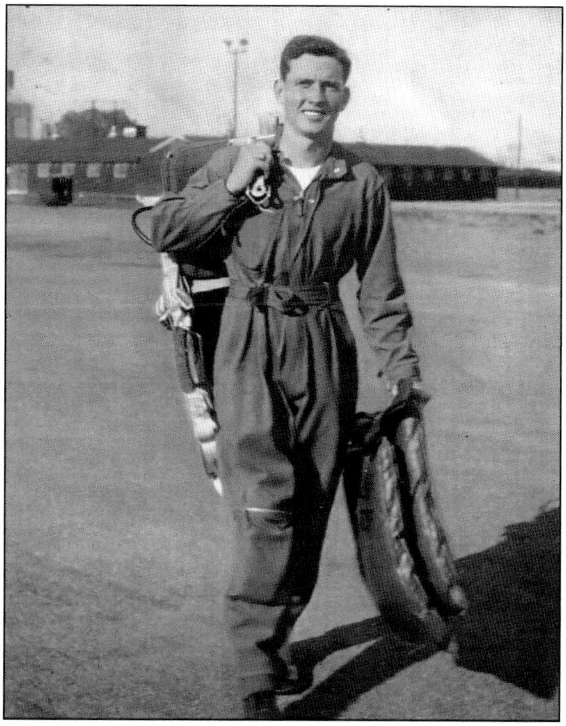

James G. Martin Jr., grandson of S.S. Lamb Jr. and Nora Lee Willis Lamb, graduated from Palmetto High School in 1942. He enlisted in the Army Air Corps, set on becoming a fighter pilot. He adored flying his P-51 Mustang, but on March 14, 1945, in his final stage of instruction, he was shot down by friendly fire in a training accident over Wiauma, Florida. His sister Lenore Stewart believes he had a subconscious premonition, because just days before his death, Martin took leave to visit several friends and relatives. (Courtesy Lenore Stewart.)

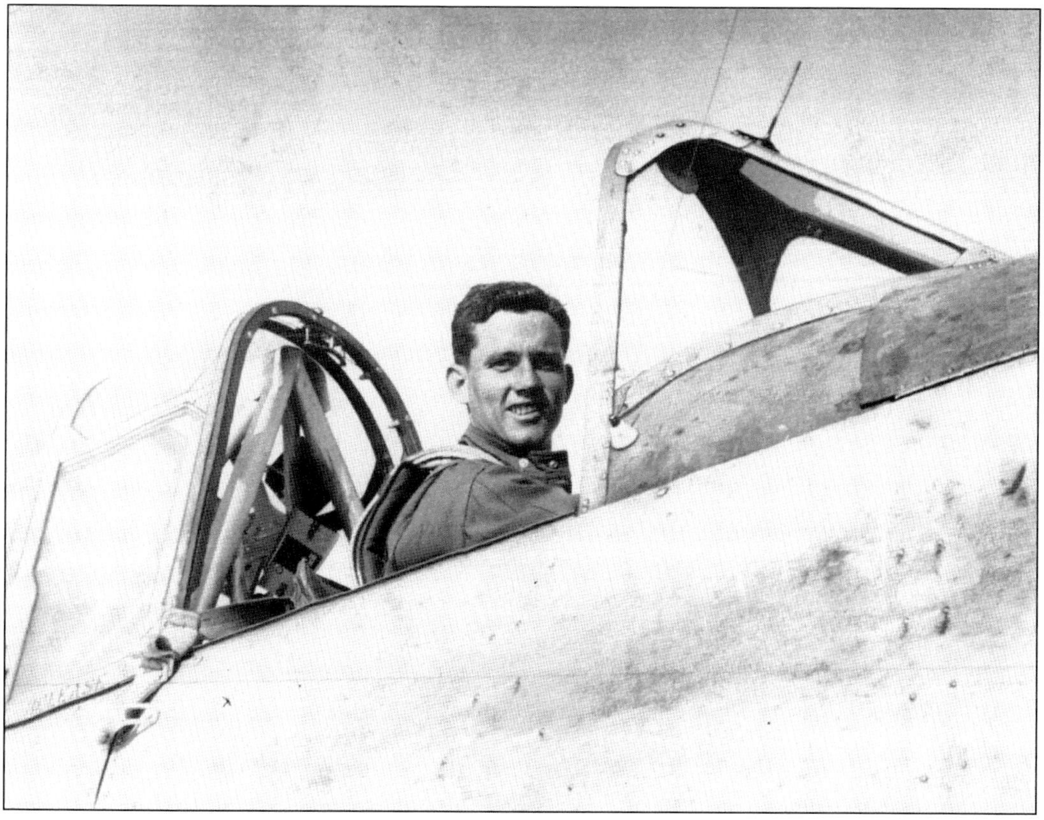

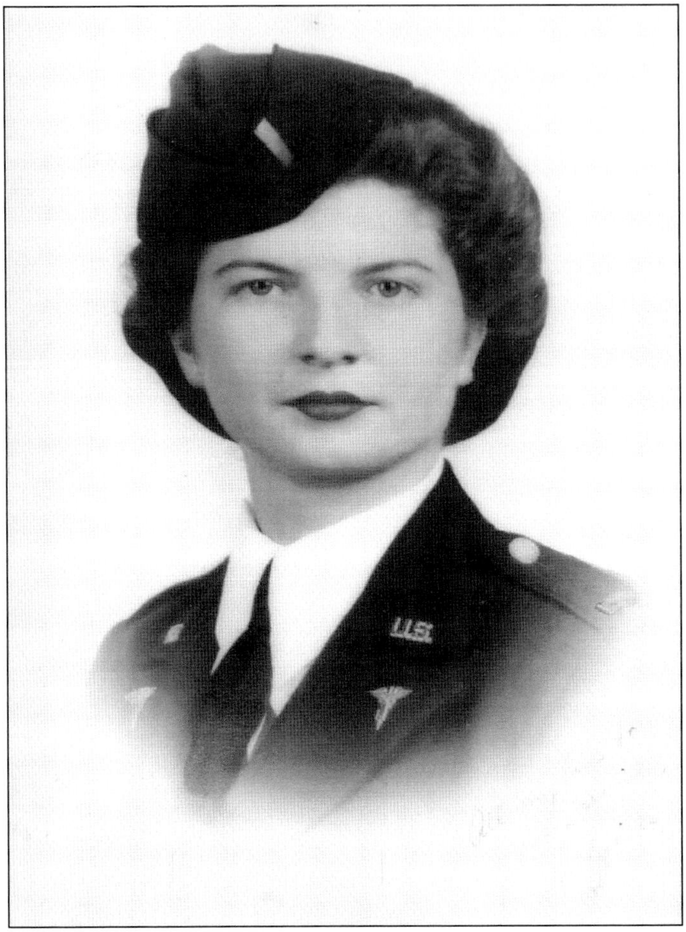

Mary Lucile Martin graduated from Palmetto High School in 1937. She served as a registered flight nurse in North Africa and the Azores during World War II. She fell in love with an Army Air Corps pilot named Seymour "Sy" Sopkin in Casablanca, Morocco. The two married under a Moroccan sunset. (Courtesy Lenore Stewart.)

The Works Progress Administration built the Palmetto National Guard Armory building in 1937. It was the first armory in Manatee County. The Florida National Guard (Company 1, 124th Infantry) occupied the building until 1962, when it was relocated to the current location on Fourteenth Avenue. During World War II, the Manatee County canning kitchen used armory facilities, and the Red Cross held training programs there. (PHC.)

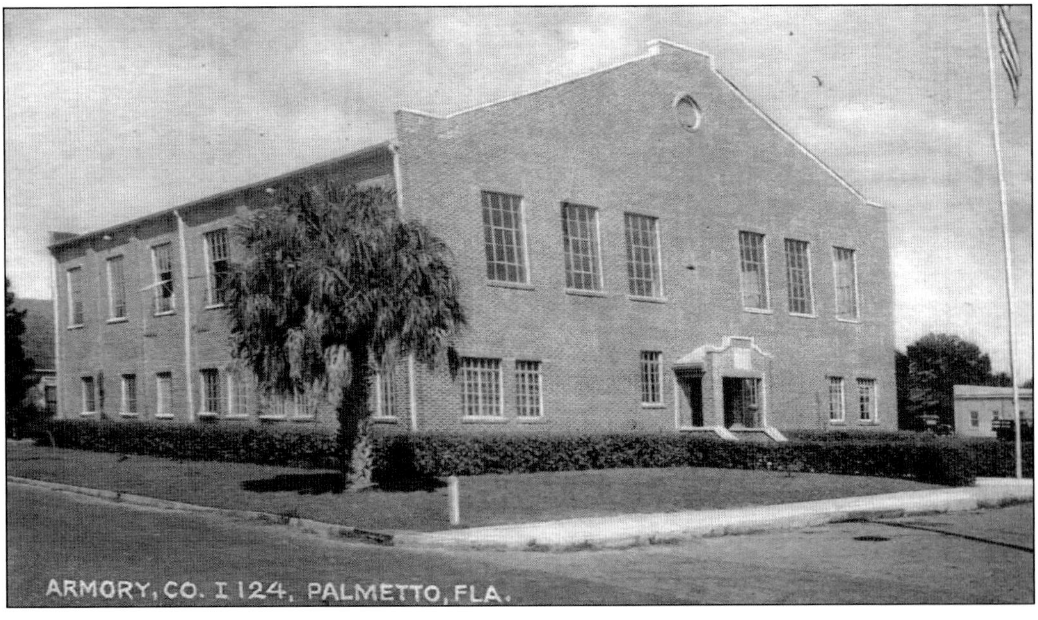

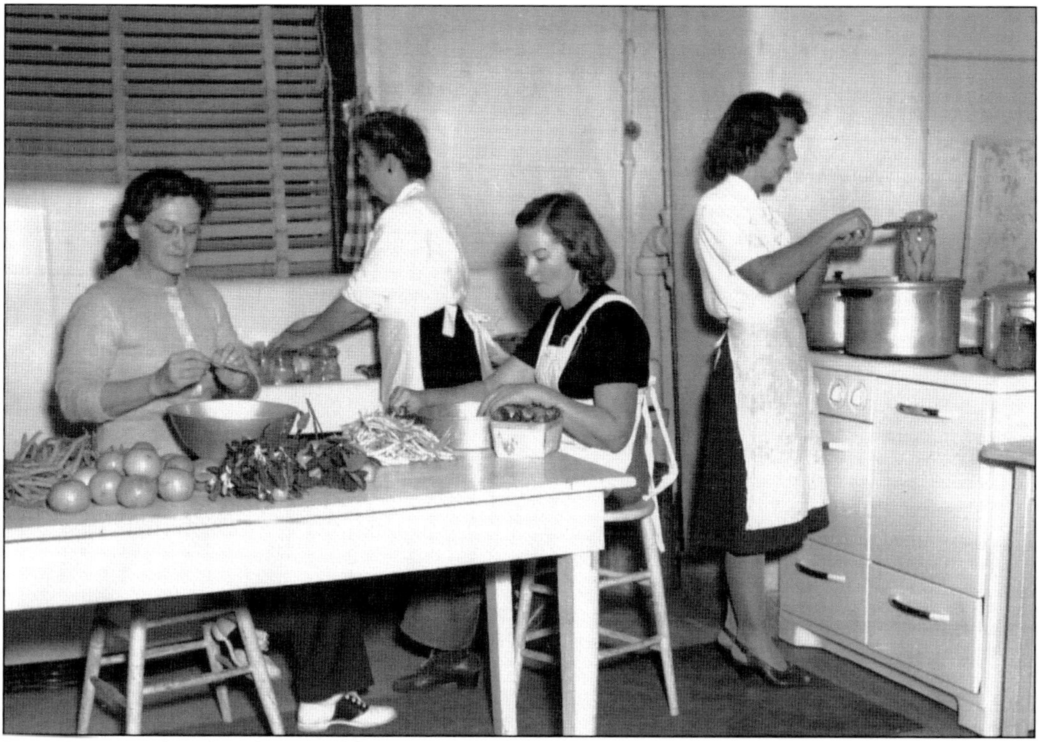

In a photograph taken by Harold R. Smith during World War II, several unidentified women take advantage of a "community kitchen" set up by Manatee County in the Palmetto Armory. Locals used the equipment to preserve homegrown fruits and vegetables by canning or jarring them. The government encouraged everyone to plant "victory gardens" in their yards to provide undersupplied families with extra food during rationing. (MCHS.)

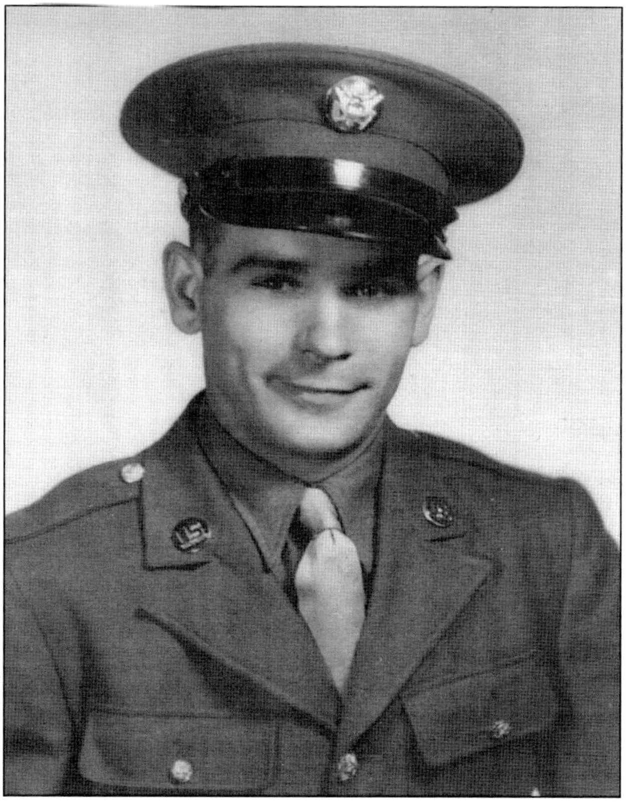

John J. Johnson Jr. received a victory medal for his World War II service. While in Europe, he was assigned to aid survivors of Auschwitz, a Nazi death camp in Poland. He took three snapshots that he donated to the Palmetto Carnegie Library. He wrote, "To see the bodies piled up, to see the way they were buried—they were bulldozed! I want to be sure the youth of today do not forget." (PHC.)

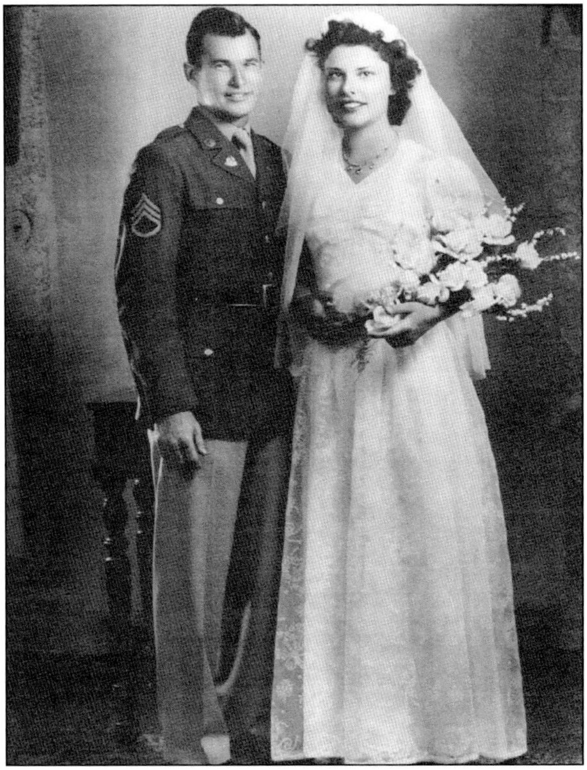

In 1939, Pres. Franklin D. Roosevelt provided funds to state National Guard units for equipment and arms as defense against the prospect of war with Germany. Young men in Palmetto joined the revamped guard, never expecting to be deployed overseas. When most of Florida's National Guard troops were called to national service on November 25, 1940, couples rushed to the chapel to wed before their sweethearts were shipped overseas. Ben Franklin Sutton and Alice Irene Anderson, seen at left, wed on April 18, 1943. Sutton, an Army staff sergeant, was wounded in Belgium at the Battle of the Bulge. Laurence Williams and Margie Anderson are pictured below on their wedding day, June 7, 1942. Williams was captured by the Germans and held at a prison camp for months, surviving on potato-peel soup. The Russians eventually liberated him. (PHC.)

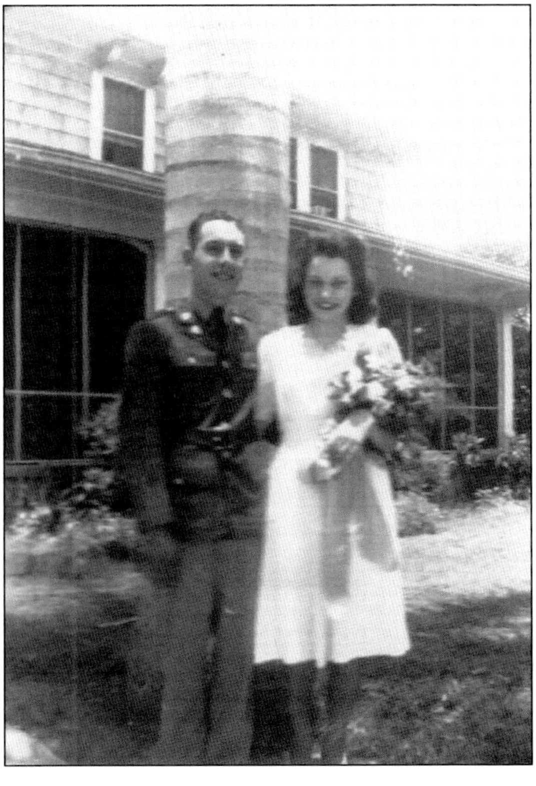

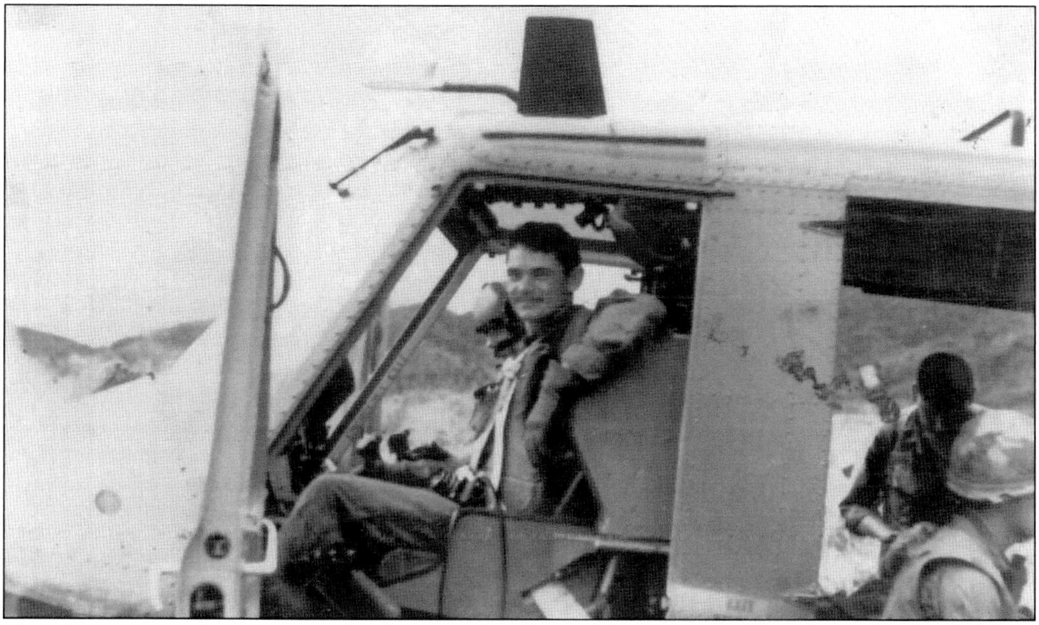

Chief Warrant Officer Ben Fredrick Sutton, son of Ben and Alice Sutton, died in combat when his helicopter crashed on September 3, 1969. He was 23 years old. Sutton was the only member of the armed forces to die in Vietnam whose home of record was Palmetto. In 1982, the Palmetto Junior Women's Club named Sutton Park in his honor and dedicated it to all veterans. Sutton represented all those who gave their lives to preserve Americans' heritage. (PHC.)

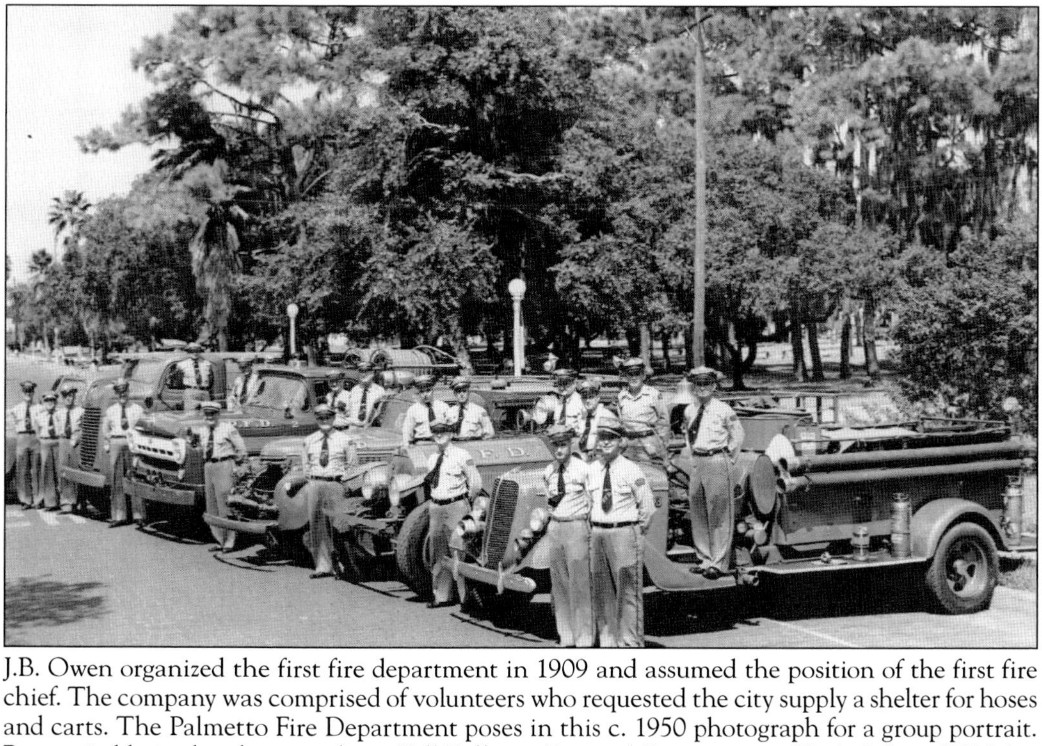

J.B. Owen organized the first fire department in 1909 and assumed the position of the first fire chief. The company was comprised of volunteers who requested the city supply a shelter for hoses and carts. The Palmetto Fire Department poses in this c. 1950 photograph for a group portrait. Recognizable in the photograph are Bill Bellamy, Bernard Simmons, and Jack Aiken. (PHC.)

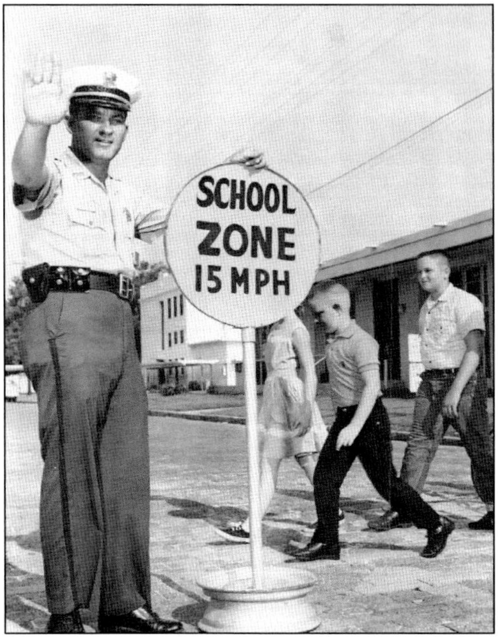

John Wright became Palmetto's first town marshal in 1893. Besides his police duties, Wright served as home-health inspector and tax collector. The first town jail was an 8-foot-by-10-foot wooden structure built on stilts over the water between Twelfth and Thirteenth Avenues. Called "Greco's Hotel" in dubious honor of its first inmate, its accommodations included a corn-husk mattress. A prisoner escaped by using the corn husks to burn a hole in the building; he then jumped through the flames into the river. In 1921, the Palmetto State Bank loaned the police department money to purchase a motorcycle under an agreement that fines collected from speeding tickets would be applied to the debt. These two photographs were taken in the 1950s. At left, Officer Bill Davis escorts students attending Palmetto Elementary School. Below, the police department poses for a portrait with Chief Al Redmond. (PHC.)

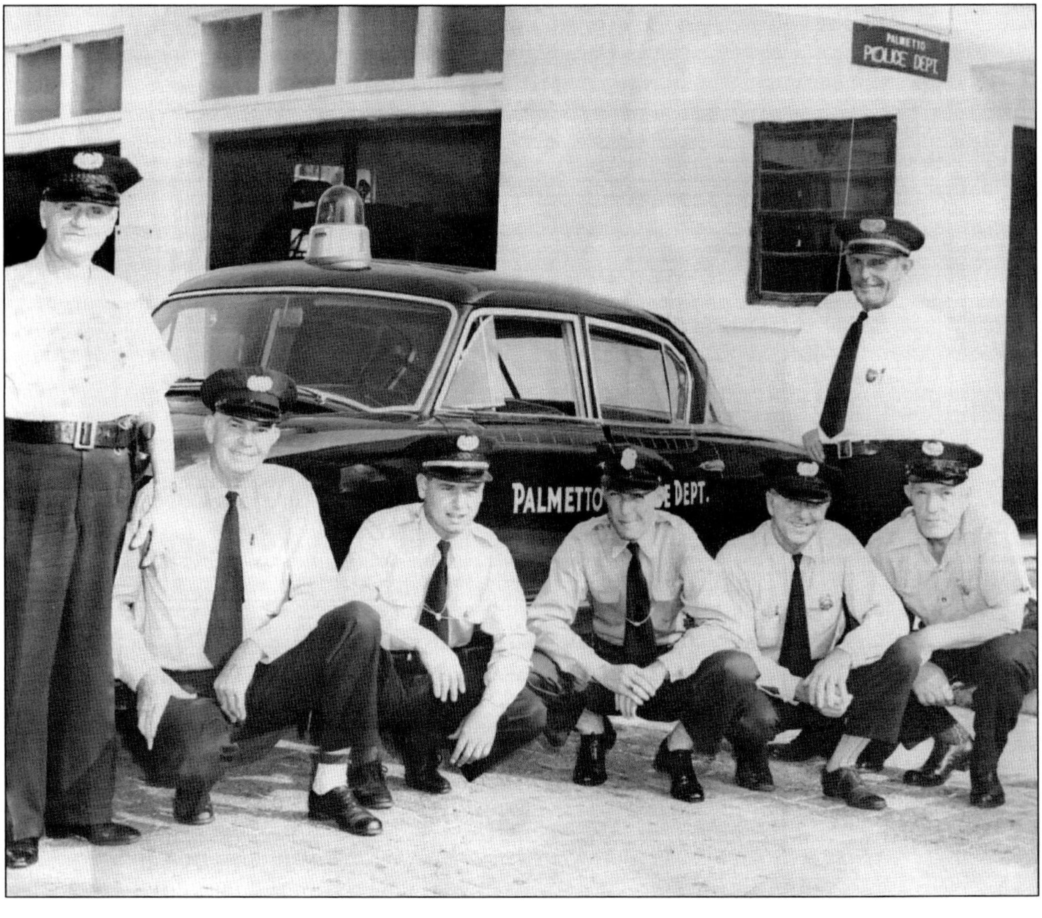

Six

Life and Times on the North Side

The social life of settlers centered on church. Piling the family in the horse-drawn wagon on a Sunday for a journey to town provided a pleasant interlude to the backbreaking work that filled the rest of the week. Many families lived in desolate locations, and even a 10-mile trip to church on rutted roads seemed a minor inconvenience considering the reward: a day spent fraternizing with friends, neighbors, and family. While daytime services were often long, singing hymns and nighttime revivals broke the tedium and entertained the entire family. After worship, women of the congregation unloaded brimming baskets and tubs filled with homemade specialties for a potluck dinner on the church grounds. Men would often roast a hog over an open fire.

Sundays were reserved for prayers and spiritual matters, but during the rest of the week, the settlers seldom lacked for leisure pursuits. Fishing has always been a particularly popular pastime. Sandy shores and tidal pools along the river harbored a supply of fiddler crabs, sand fleas, and shrimp to use as bait. Poles were made of cane, and nets were crafted from cotton twine. The local bays and bayous held scallops, oysters, crabs, and clams. It was common for families to camp on one of the many cays and roast their catches on screens set over an open fire. Another treat was to pack a picnic lunch for boat excursions to Anna Maria, Egmont Key, or up the Manatee River to Rye Bridge or Fort Hamer, both popular swimming spots.

Isolated from town and neighbors as they were, settlers seeking camaraderie formed social clubs. The Masonic Lodge was one of the first organizations for the men of the city. Women formed the Village Improvement Association, and young men eagerly partook in Boy Scout adventures. Several of Palmetto's fraternal organizations were the first of their kind in the county.

School served both an educational and a communal role. When S.S. Lamb hired Frankie McKay (later Frankie Howze) as the first principal at the town's first school, she promptly promoted all types of activities for children. She put on school plays and organized the first sports teams in Palmetto, including football, and she ambitiously took on coaching duties. Palmetto gradually became the center of social life on the north side of the Manatee River.

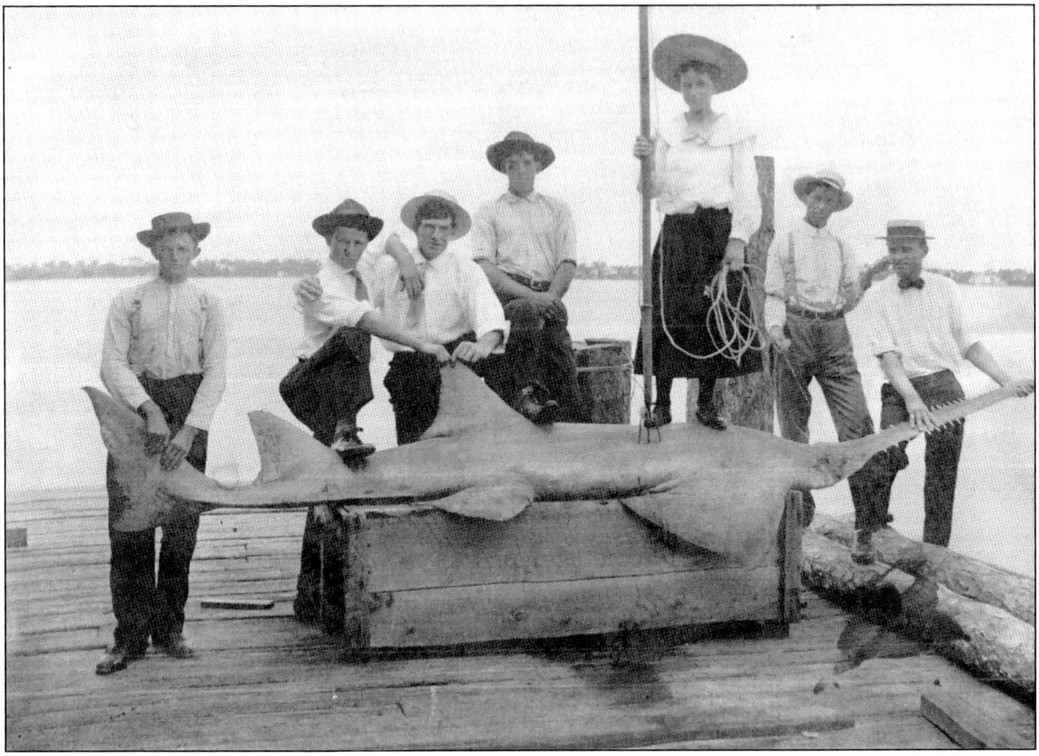

This sawfish, caught on July 7, 1907, was 14.5 feet long and weighed more than 1,000 pounds; the saw alone was 5 feet long. It was caught on Palmetto Wharf at the end of what is now Thirteenth Avenue. From left to right, Cleveland Peudorvis, Charlie Waller, Millard McDougall, Bowham Duncan, Maggie Wright Arnold, George Arnold, and Barrett Curry pose with their colossal catch. Maggie jokingly sticks a gig in the giant fish. (PHC.)

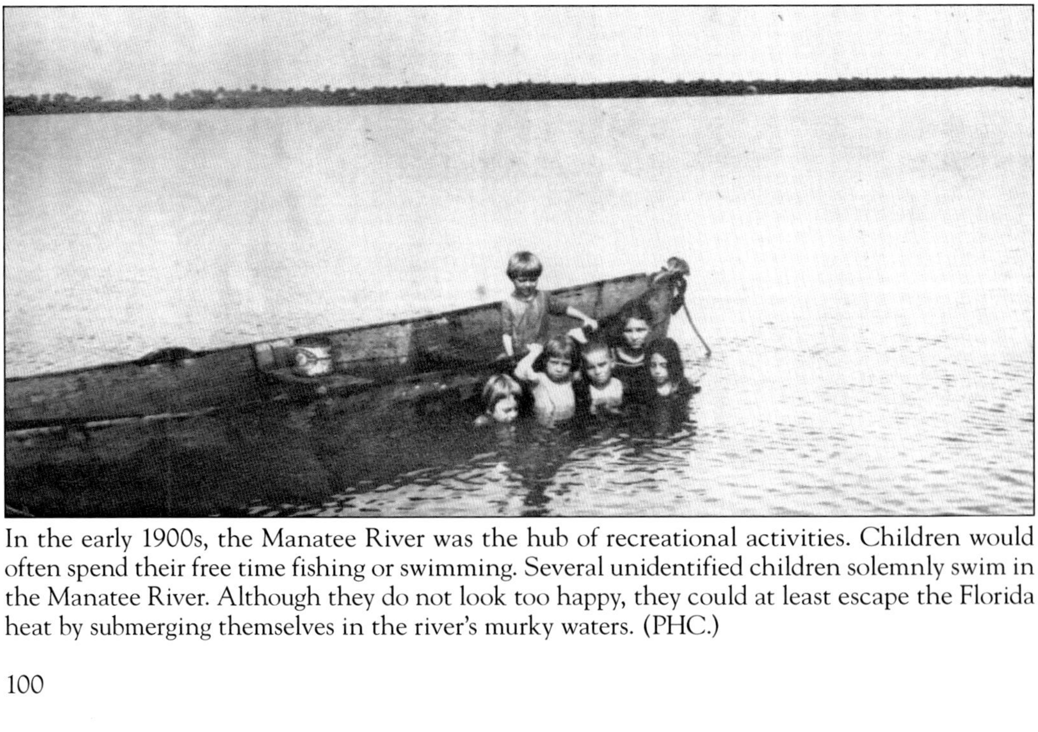

In the early 1900s, the Manatee River was the hub of recreational activities. Children would often spend their free time fishing or swimming. Several unidentified children solemnly swim in the Manatee River. Although they do not look too happy, they could at least escape the Florida heat by submerging themselves in the river's murky waters. (PHC.)

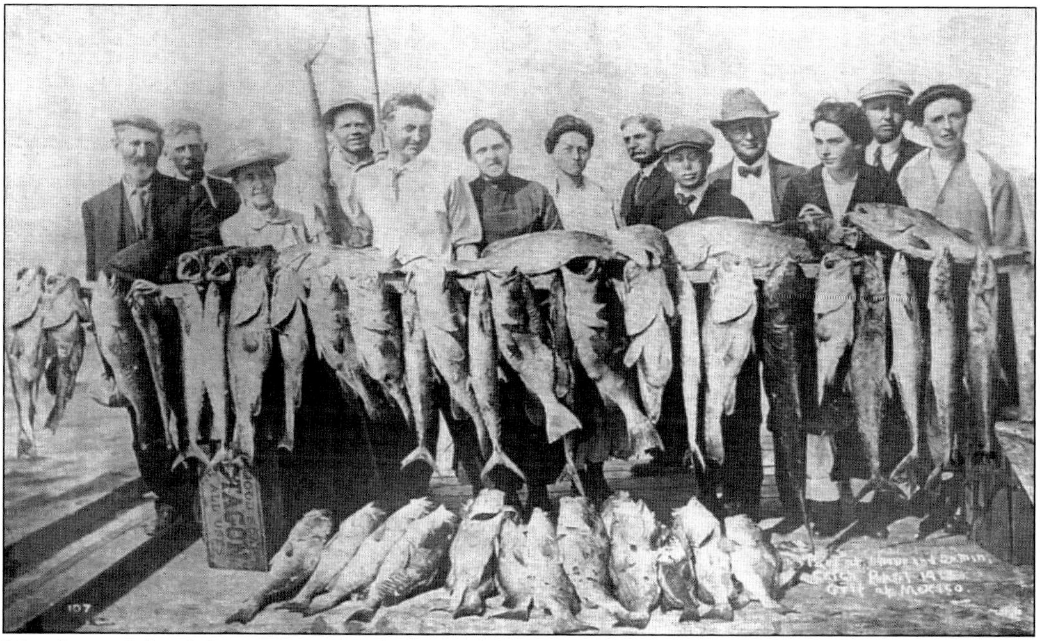

Members of this fishing party caught a good day's catch to say the least. From 1869 to 1900, the first records, species descriptions, and ecological studies were conducted in Tampa Bay. Scientists have concluded the only species to become extinct in the area since that time is the sturgeon. In the winters of 1886 and 1887, locals harvested 2,500 sturgeon that yielded over 10,000 pounds of caviar—the species never recovered. (PHC.)

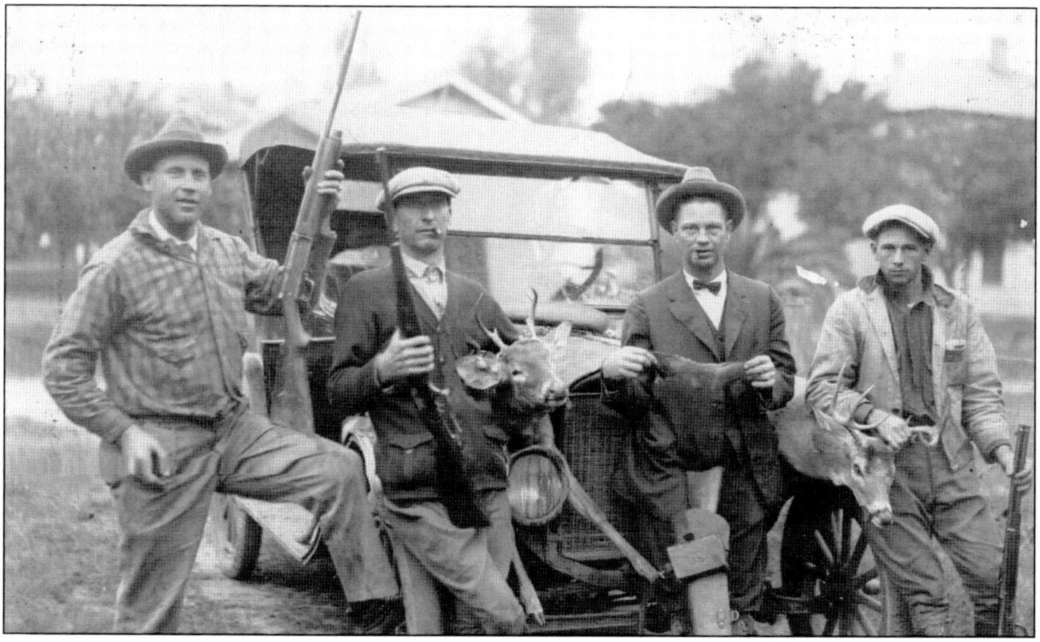

Hunting has long been another popular pastime in Palmetto. Tracking game meant exploring Florida's pristine environment, enjoyable in itself. Bringing home food for the family table was a bonus. In this c. 1920 photograph, four unidentified young men celebrate a successful day of hunting. Apparently wearing a bow tie was considered perfectly acceptable hunting attire. (PHC.)

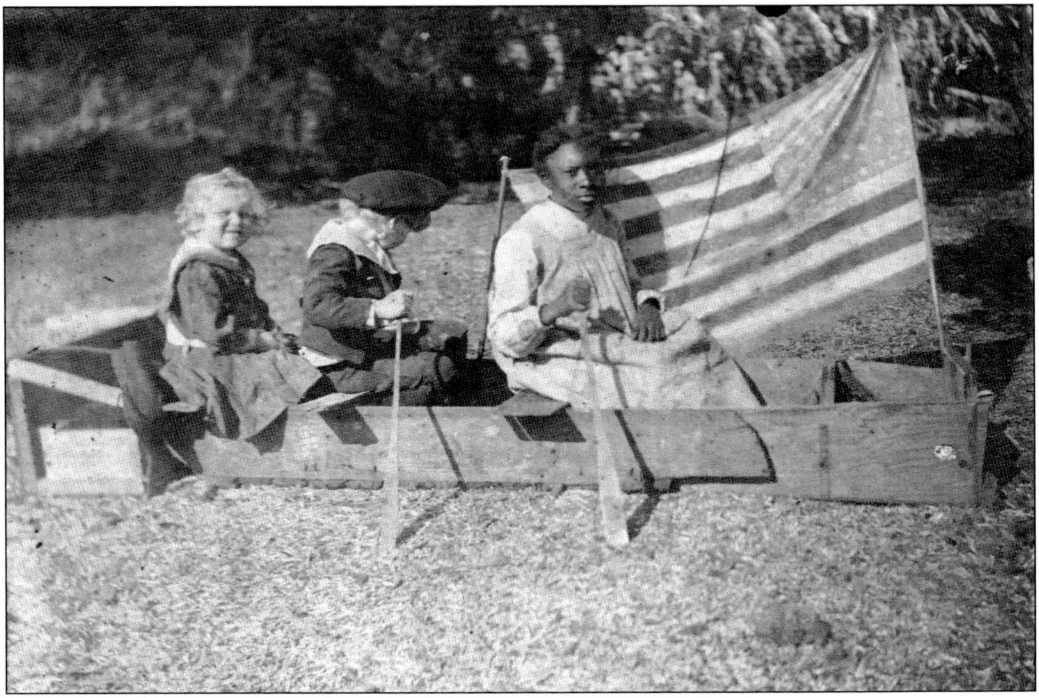

Two Howze children and an unidentified girl pretend to venture out on a makeshift vessel under an American flag sail. It is assumed that their desired destination was somewhere other than their front yard, which faced the Manatee River and its endless river traffic. Luckily, their epic voyage ended in just enough time to eat the cake that was served at the birthday party they were attending on March 10, 1907. (PHC.)

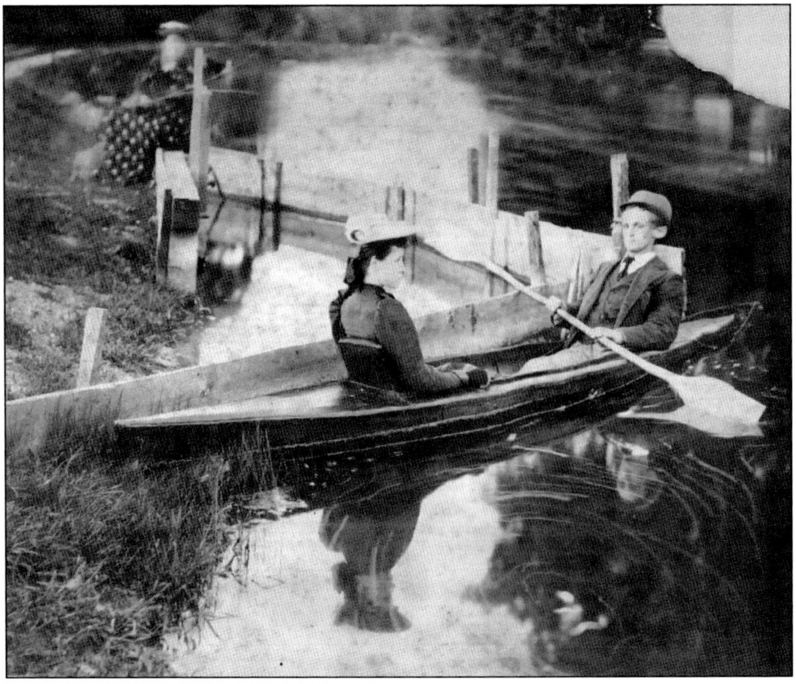

A sharp-dressed young couple takes a canoe trip to the upper Manatee River around 1900. In the background, a woman fishes on the bank. The small, sleekly designed canoe was intended for use on calm freshwater ways. Seasonal fisherman in the area still used large dugout canoes, often with sails. Their size and weight enabled them to take on the most severe ocean swells. (PHC.)

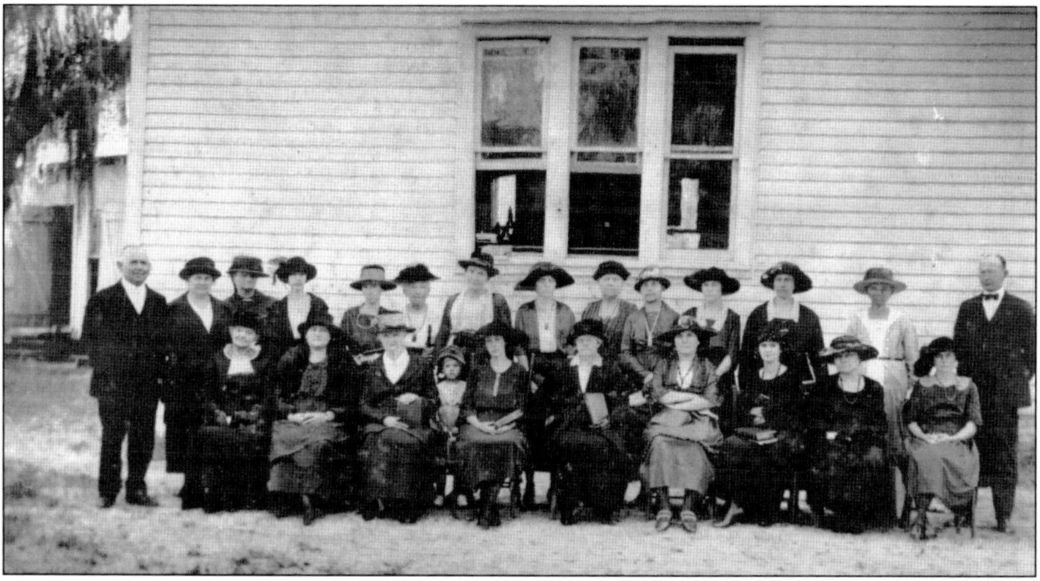

The original Baptist church was near Frog Creek. In the late 1870s, it was relocated to Gillett and named Benevolence Baptist Church. After it was destroyed by fire, the congregation elected to rebuild in Palmetto. Palmetto Baptist Church was organized on January 5, 1892, with 22 members. In 1987, S.S. Lamb donated land for a wooden structure. Members of "the best Baptist church class" pose in front of the old church in this 1920s photograph. (PHC.)

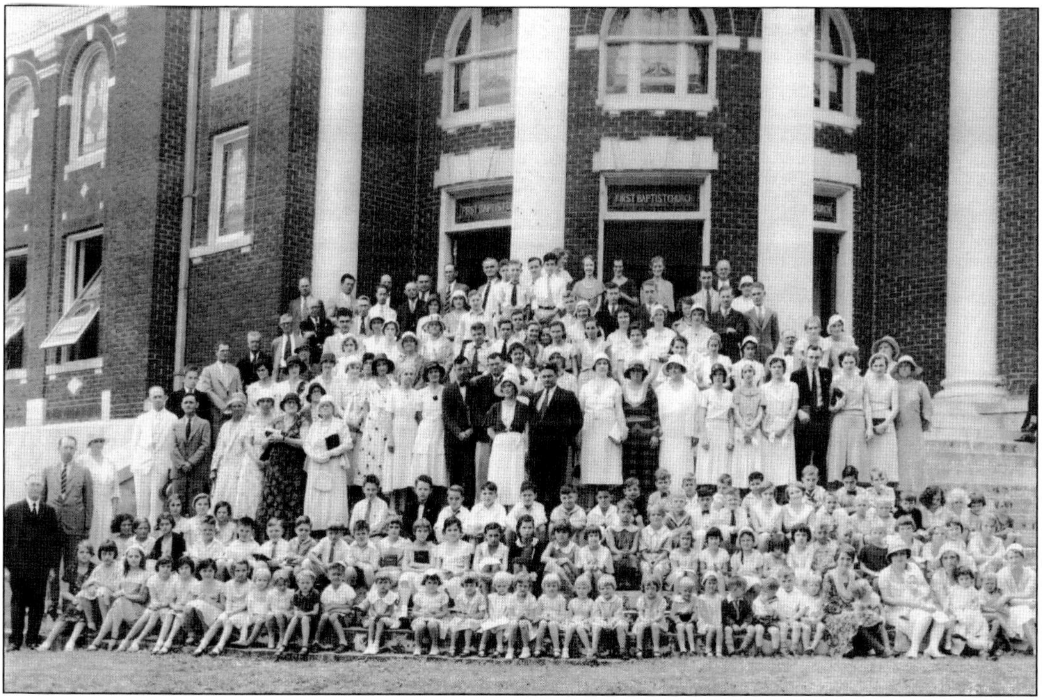

Sarah Lamb donated the property on which the Baptist church stands today. In 1919, construction began on a brick edifice that cost $30,000. In 1941, the church incorporated as the First Baptist Church in Palmetto, the name it maintains to this day. The whole congregation stands on the steps of the brick structure. (PHC.)

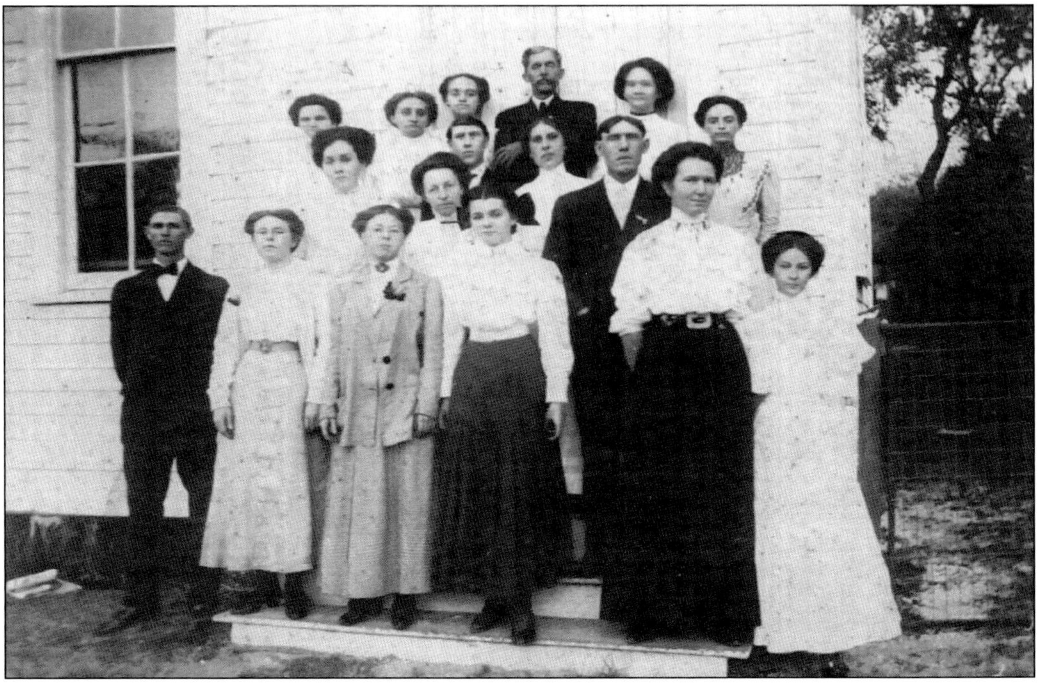

The first Methodist church was in Manatee. Locals crossed the river by boat to attend services. The church became known for the exhilarating revivals held on moonlit nights. In 1892, a Palmetto branch began operating out of the school building. S.S. Lamb donated land on Eighth Street for a church; the property sold in the 1920s. Mrs. S.S. Lamb donated the current lot for the new brick building. (PHC.)

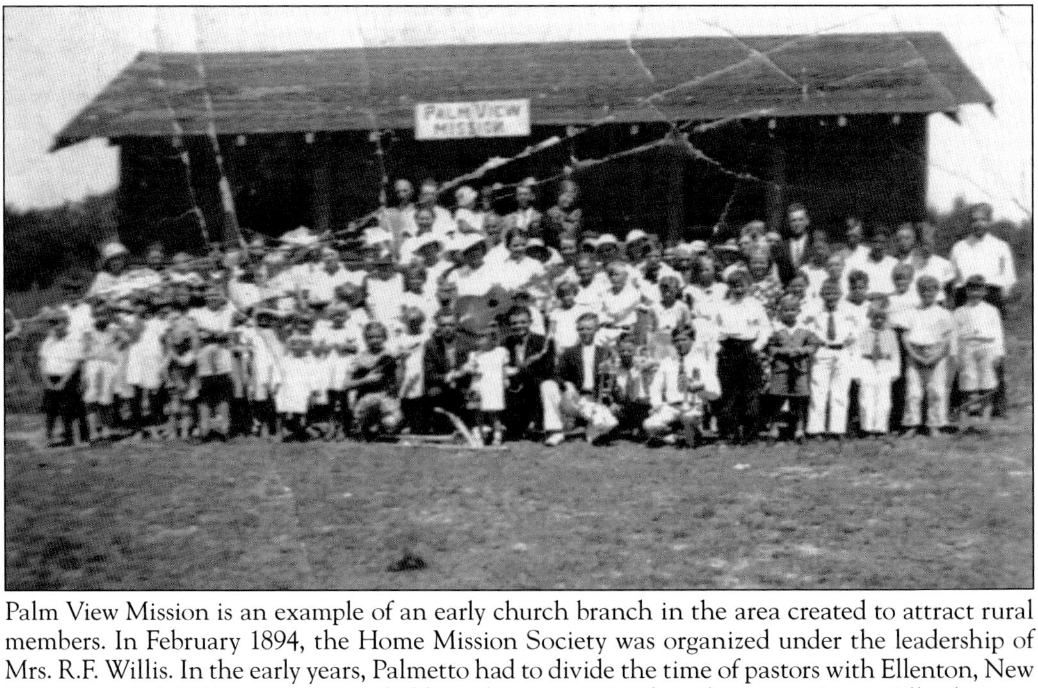

Palm View Mission is an example of an early church branch in the area created to attract rural members. In February 1894, the Home Mission Society was organized under the leadership of Mrs. R.F. Willis. In the early years, Palmetto had to divide the time of pastors with Ellenton, New Hope, and Oak Hill. Services were held on alternating Sundays. (Courtesy James Allen.)

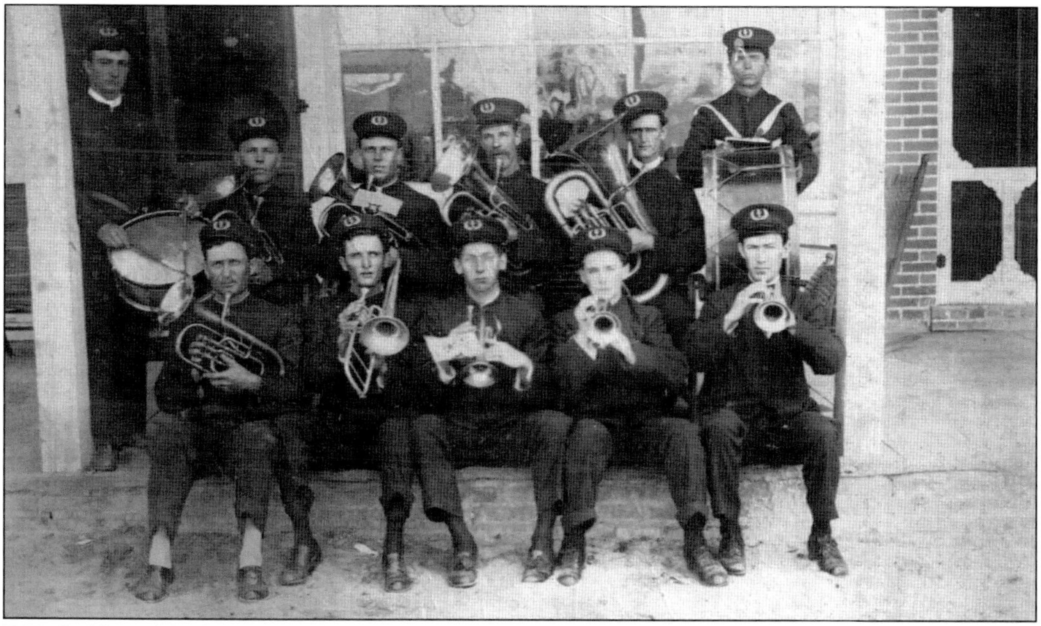

R.E. Cobb and Earl Mason, who played bass horn, led the Palmetto City Band. John Pope Harllee and two of the Nettles brothers were members in 1910. One winter evening, after a concert at Warren's Opera House, the men stayed in a hotel because the Palmetto ferry was closed. With no heat and few blankets, a fight broke out over who would get the covers. (PHC.)

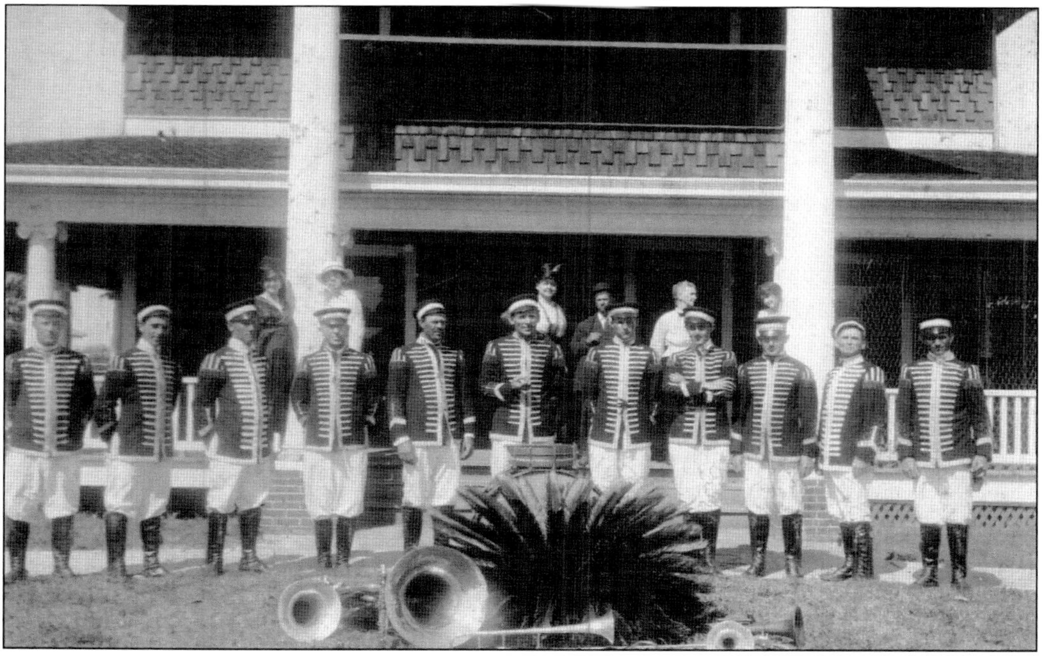

Members of the Phillips Band pose outside the Phillips's home at 1316 Fourth Street West. C.A. "Cap" Philips was the owner, lead actor, band member, and playwright for the Phillips Dramatic Company. He convinced the railroad to build a spur behind his property for easy access to his show cars. Hudson Colwell, Mr. Broiler, and Mrs. Harry Grant were actors who came to Palmetto to perform with the company. (PHC.)

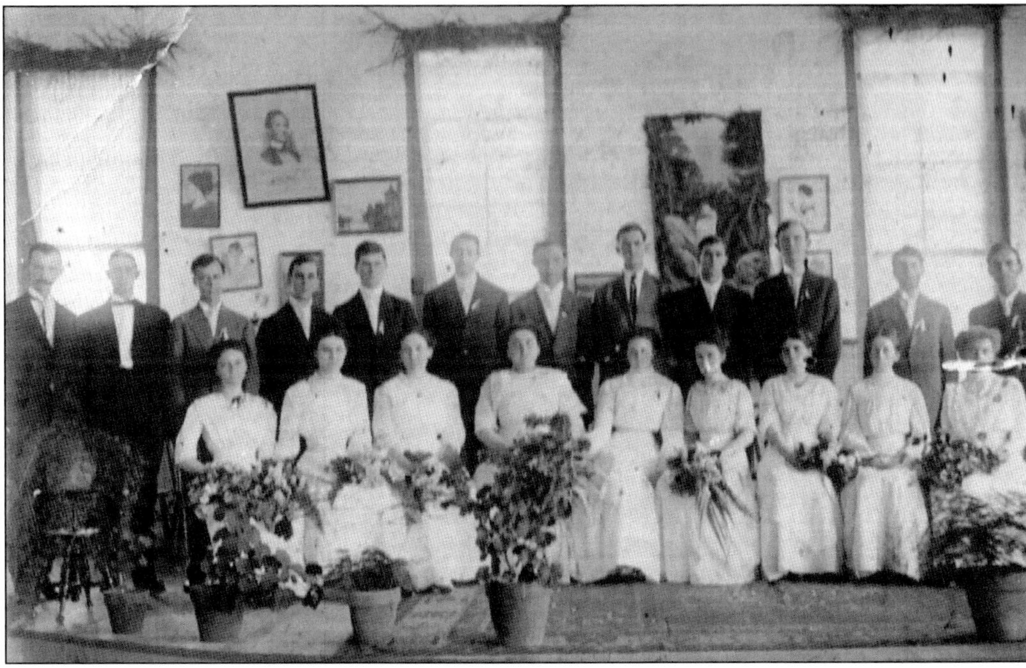

One of the earliest organizations in Palmetto was the Independent Order of Odd Fellows. It was a fraternal association derived from a similar 18th-century British service society. The "brothers and sisters" of the club sought to welcome travelers and help those in need. "Visit the sick, relieve the distressed, bury the dead and educate the orphan" was the organization's motto. (PHC.)

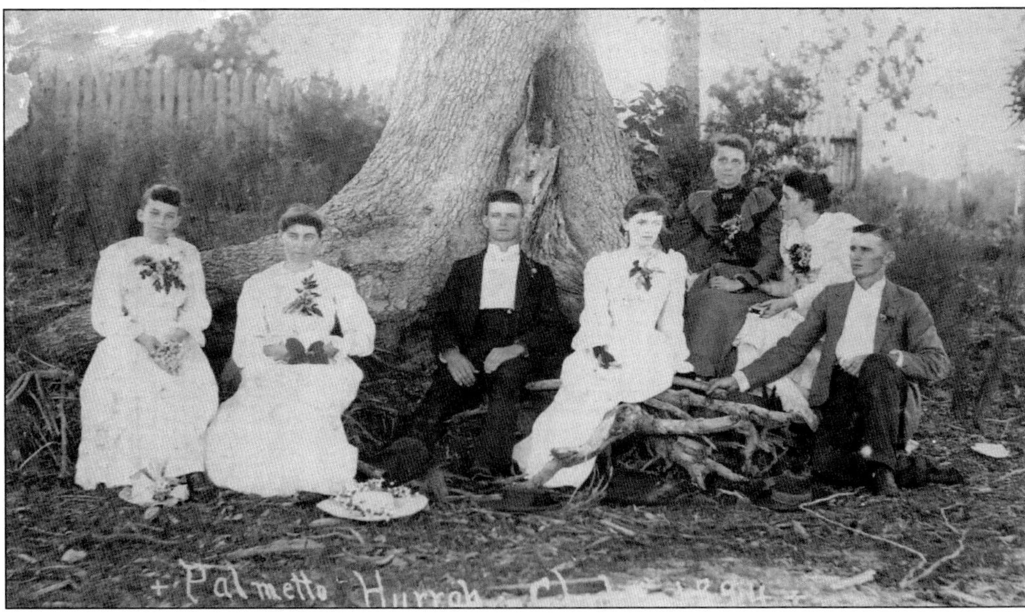

The Hurrah Club was comprised of young men and women who met at various locations to socialize. Members enjoyed getting to know one another without the pressure of committing to a "date," and parents were more comfortable with the arrangement. Among those in this 1896 photograph are Nora Lee Willis, Martha "Mattie" Lamb, S.S. Lamb Jr., Laura "Lollie" Lamb, and Albert Haley. (Courtesy Lenore Stewart.)

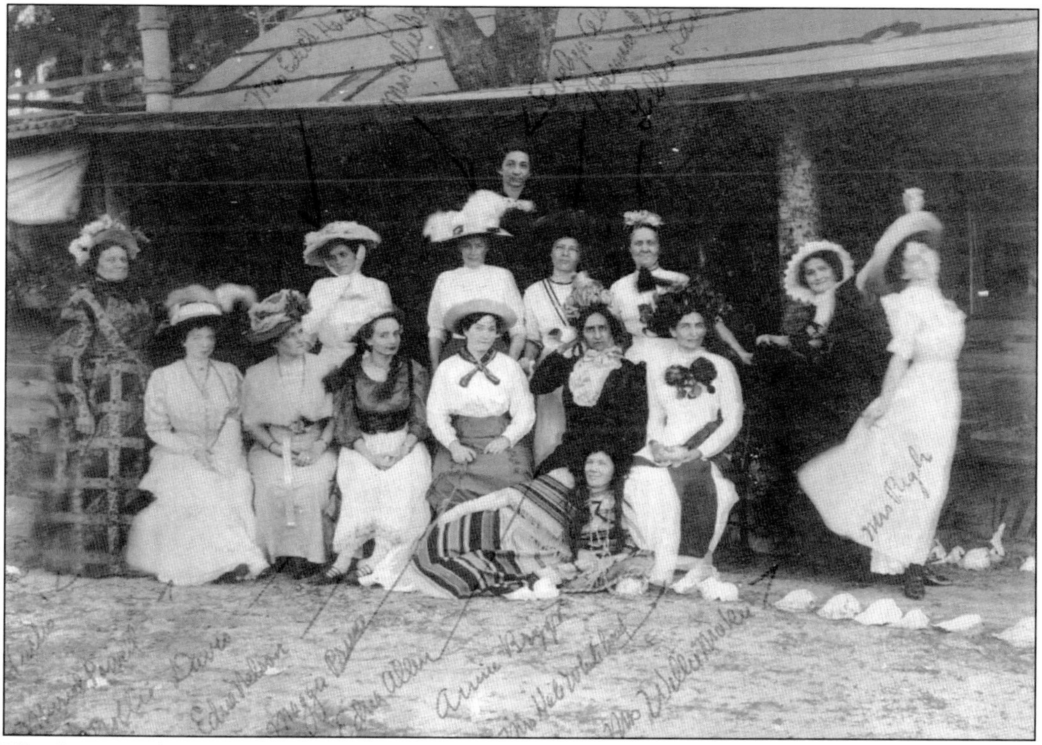

The first members of the Palmetto Women's Club, then called the Village Improvement Association, pose in their "old maid" costumes for a skit held about 1900. Recognizable in the photograph are Annie Biggs, Mrs. Davis, Edna, Mrs. Futucker, Mrs. Howze, Mrs. Clark, Mrs. Waller, Mrs. Lamb, Mrs. Willie McRee, and Mrs. Pugh. Mrs. Allen (back) was the director. (PHC.)

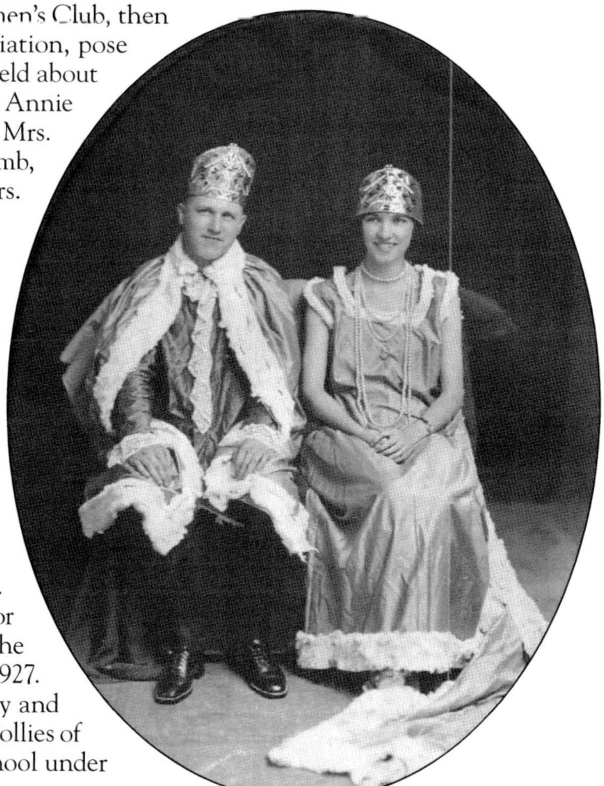

The Follies was a widely supported musical revue whose profits benefited the Palmetto Women's Club construction. Participants included local townspeople or students who enrolled for class credit. The Follies ran for three years, from 1924 to 1927. In this photograph, King Pete McGinley and Queen Louise Mathis are elected at the Follies of 1924, held at Palmetto Intermediate School under the direction of M.H. Dorsett. (PHC.)

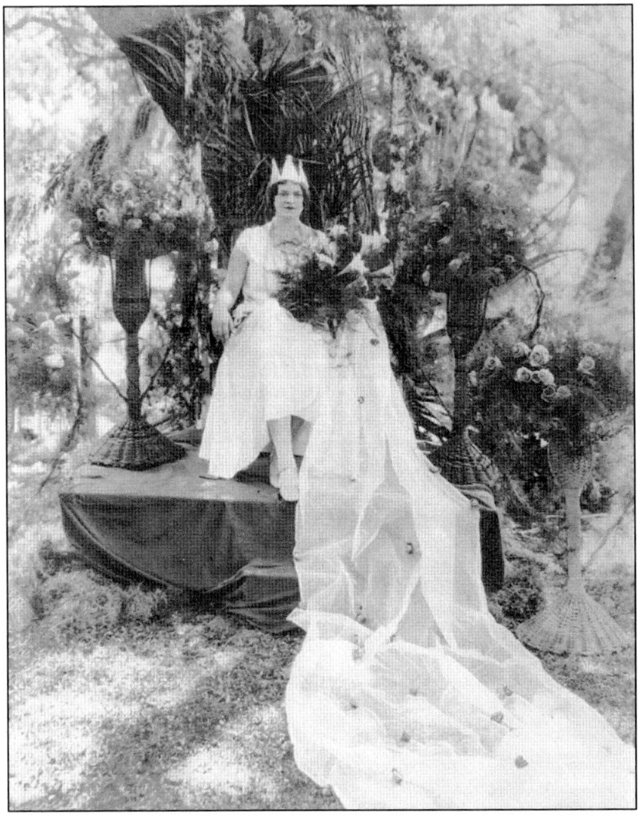

Every year, Palmetto held an annual tomato festival, and an attractive young woman was elected as tomato queen. Queen Mary Turner sits upon her throne of palm fronds, as bouquets of roses surround her royal highness. (PHC.)

Manatee Fruit Company hosted an annual Mayday picnic, where local ladies performed a play for guests. The recognizable actors are Lula Haymond, Mary Young, Sam Tyler Rowlett, Annie Lamb, Clara Booth, Eva James Weatherall, Eva Albritton, and Callie Robinson. (PHC.)

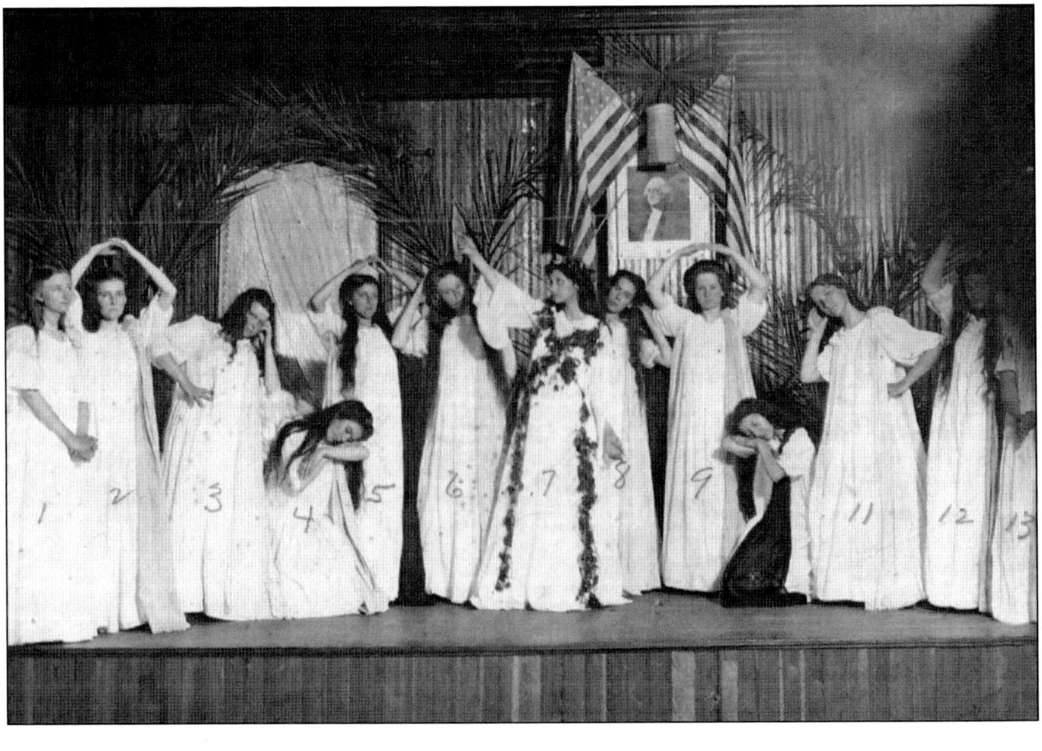

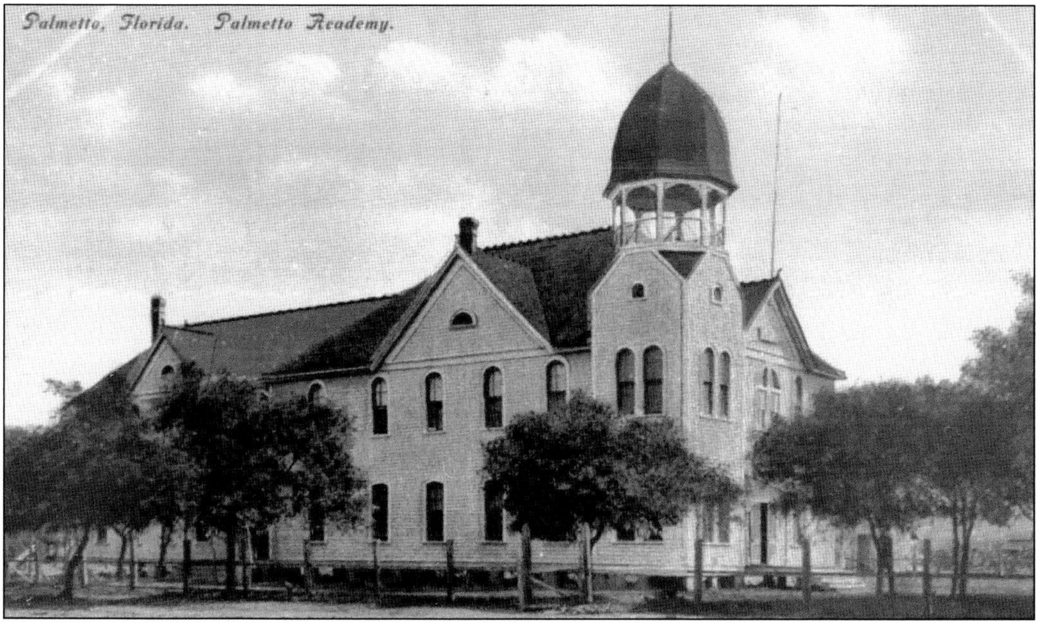

Palmetto Academy was the city's first publicly funded school and the primary reason for the town's incorporation. Prior to its construction in 1895, classes had first been held in the residence of S.S. Lamb, then in Madame Joe's old cabin, and finally in a multi-use, public-gathering structure that stray animals sometimes occupied. Frankie McKay, an early leader in education, was reportedly appalled by its condition when she arrived in 1889. (PHC.)

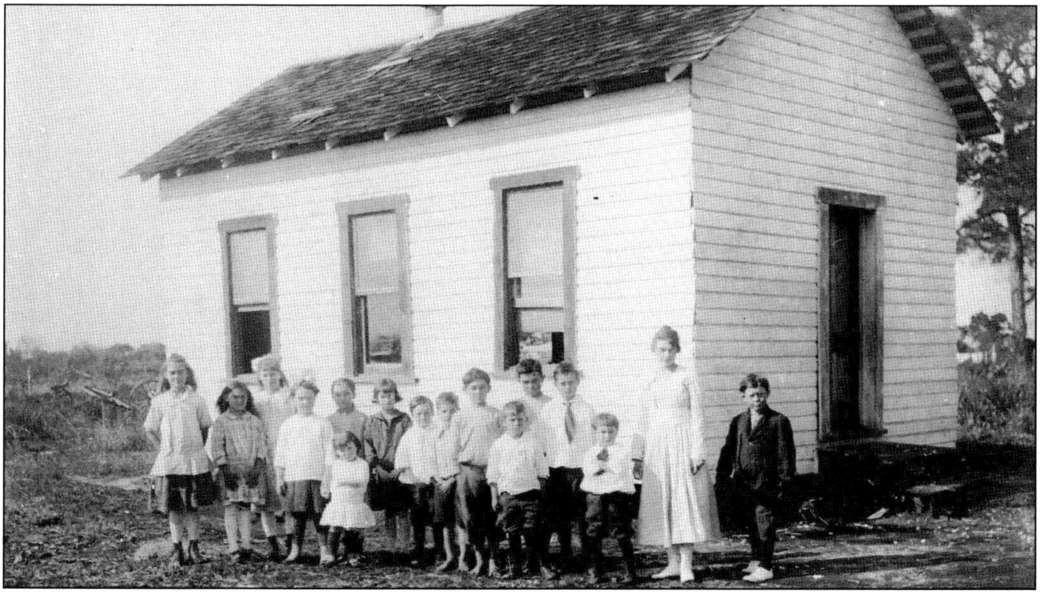

Despite a violent fatality at the location, construction of the one-room school on Snead Island was completed around 1910. Capt. Edward Pillsbury first used the washhouse next to his home as a classroom for island children. J.H. Judd donated property for a new school, but the postmaster, Mr. Hosletin, was unhappy with the site. During construction, Judd became so disgruntled he came after Hosletin with a gun, but Hosletin shot and killed Judd in defense. Classes started shortly after the incident. (MCHS.)

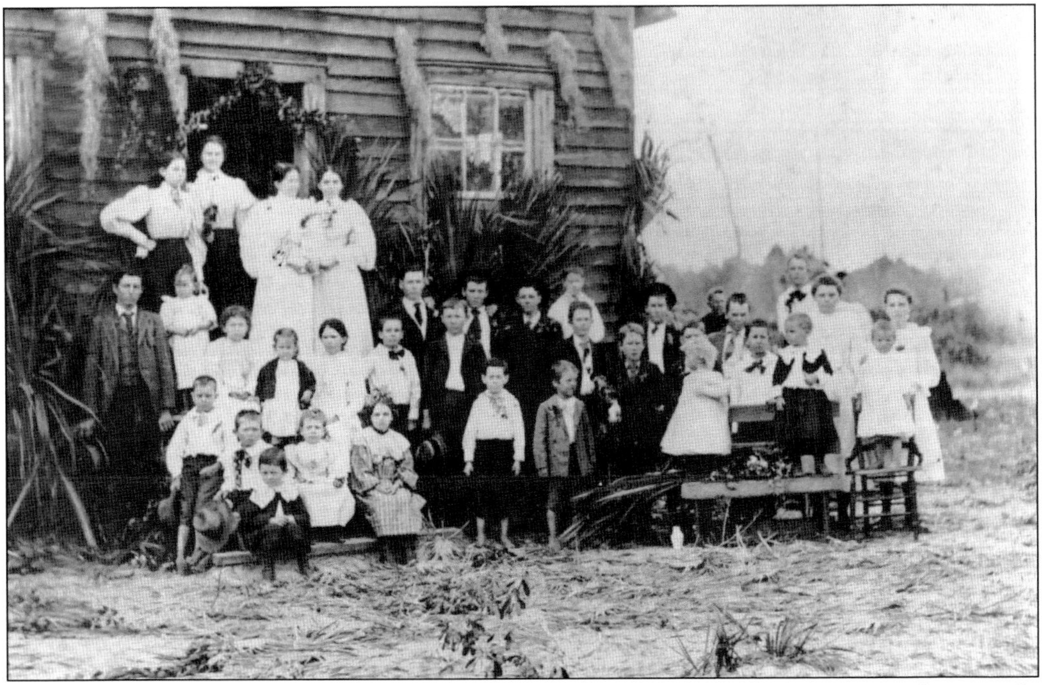

The Seventh Day Adventist Church in Palm View doubled as a school. In this 1902 photograph, Louise Horn poses with her students. She received $13 a month as her salary. She later married Ed Andress of Snead Island. (PHC.)

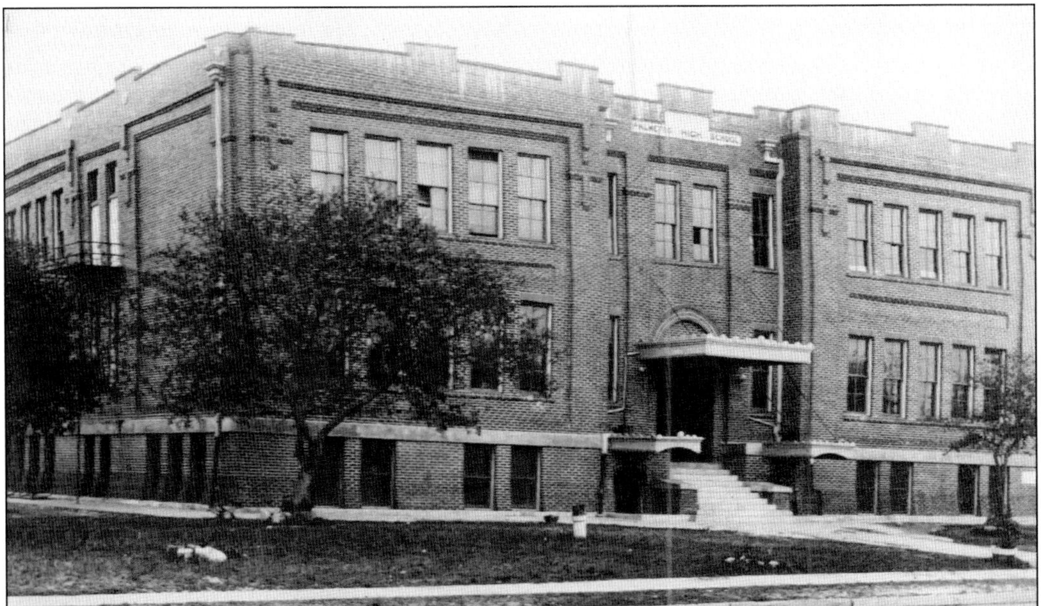

The brick school was constructed in 1912 along Tenth Avenue between Seventh and Eighth Streets. It had two stories and a basement that housed the furnace and physical education facilities. Classrooms lined the first floor, and a small auditorium and stage occupied the second level. With the addition of another building in 1923, the brick building became Palmetto High School. It was used until its demolition in 1959. (PHC.)

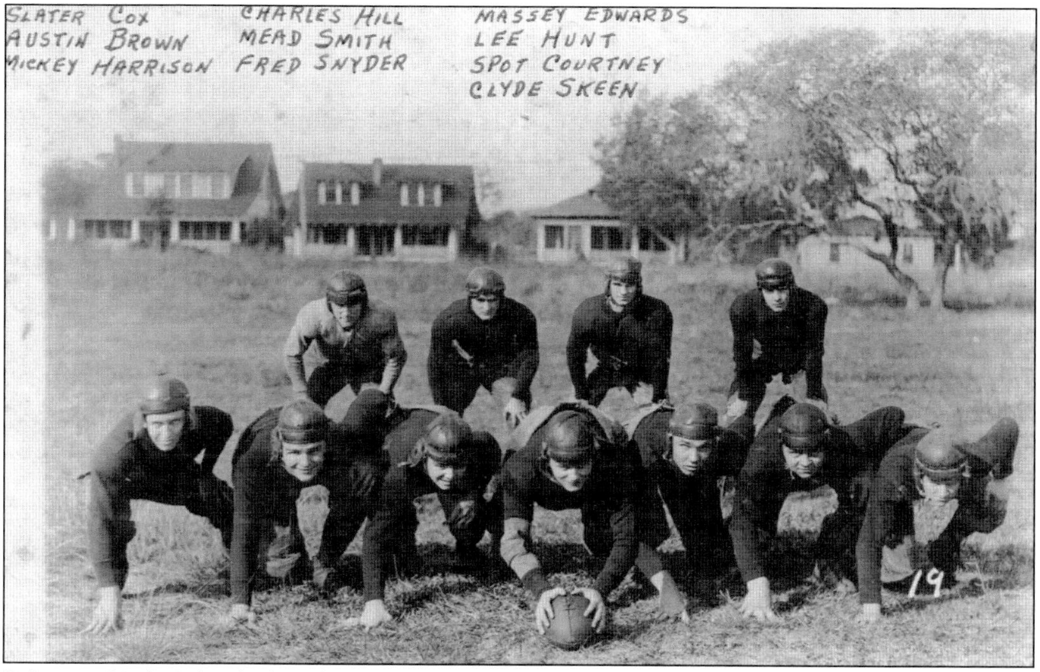

Palmetto High School's first football coach was Frankie Howze. Sutton Park was the site of the original football field, which earned the title of "Sandspur Stadium" because it was overrun by the prickly weed. The house to the far left is the Jack Travis home on Twelfth Avenue West. The photograph lists 10 of the 11 players seen here. (PHC.)

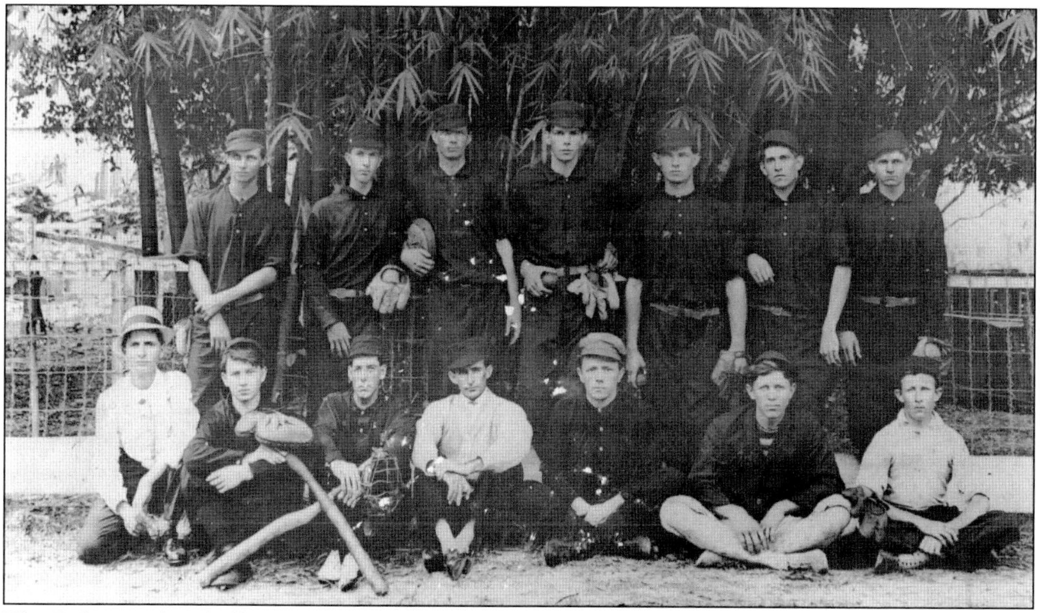

The Palmetto baseball team was organized in 1908, and it often played a team from Terra Ceia. One of its most anticipated games occurred in 1910 when players from both local teams were selected to play against a US Army squad at Egmont Key. In 1924, the Palmetto City Council purchased property for the construction of a professional stadium. The Buffalo Bisons played in Palmetto Stadium from 1925 to 1929. (PHC.)

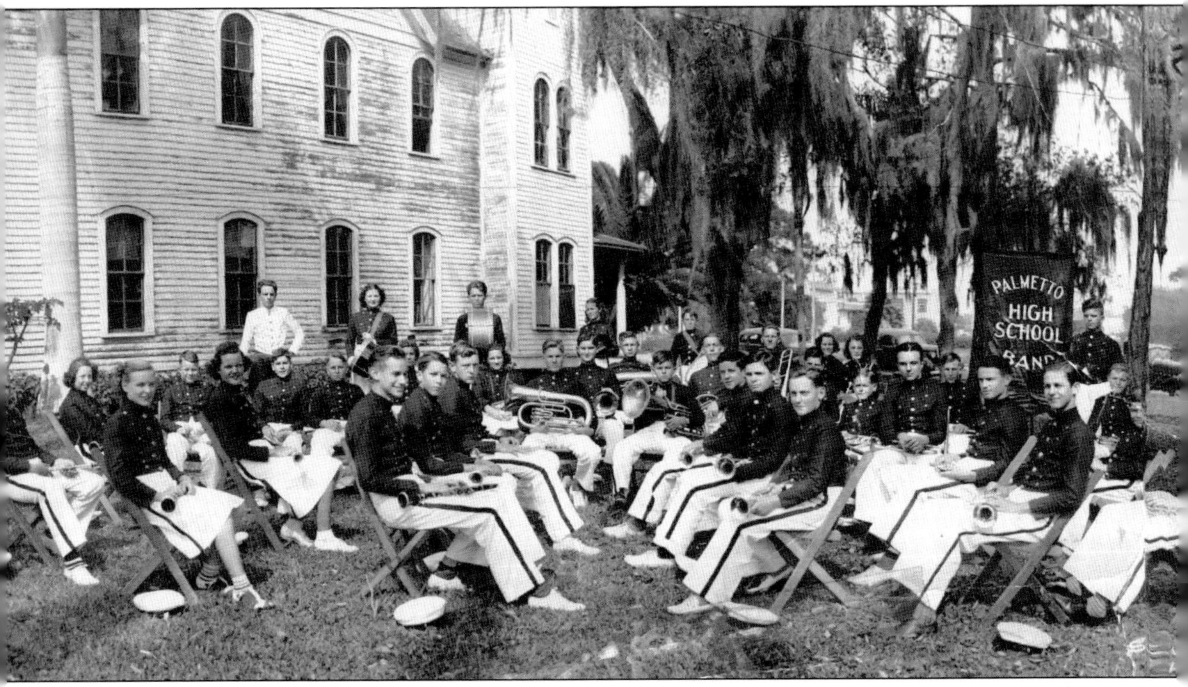

The Palmetto High School Band is seen here just before a performance in 1939. Its members in the front rows are, from front to back starting on the left, (first row) M.K. Wheder and Peggy McLeon; (second row) Ann Bennett and Mary Martha Mathews; (third row) Kenneth Parrish, Dan McClure, and George Harrison; (fourth row) Jack Hayes, Howard Thompson, and Jimmie Martin; (fifth row) Roy Le Grange, Lewis Hamilton, Wade Dix, and Buck Toft; (sixth row) Jimmy Moore, Dick Dungan, and Enoch Wingert. The members seated across the back are, from left to right, S.W. "Sam" Miley, Orrin Christian, Bill Snyder, Nalda Fugus, Katherine Garcia, Charles Thomas, W.D. Bell, Kendall Owen, Jimmy Caldwell, Jimmy Duke, Kenneth Garcia, Ouida Styers, Dot Styers, and Bobby Wilson. The members standing in the back are, from left to right, Carey Proctor, Vivienne Haymen, Albert Flowers, Dolores McRee, Calvert Courtney, and James Fred Bell. (Courtesy Lenore Stewart.)

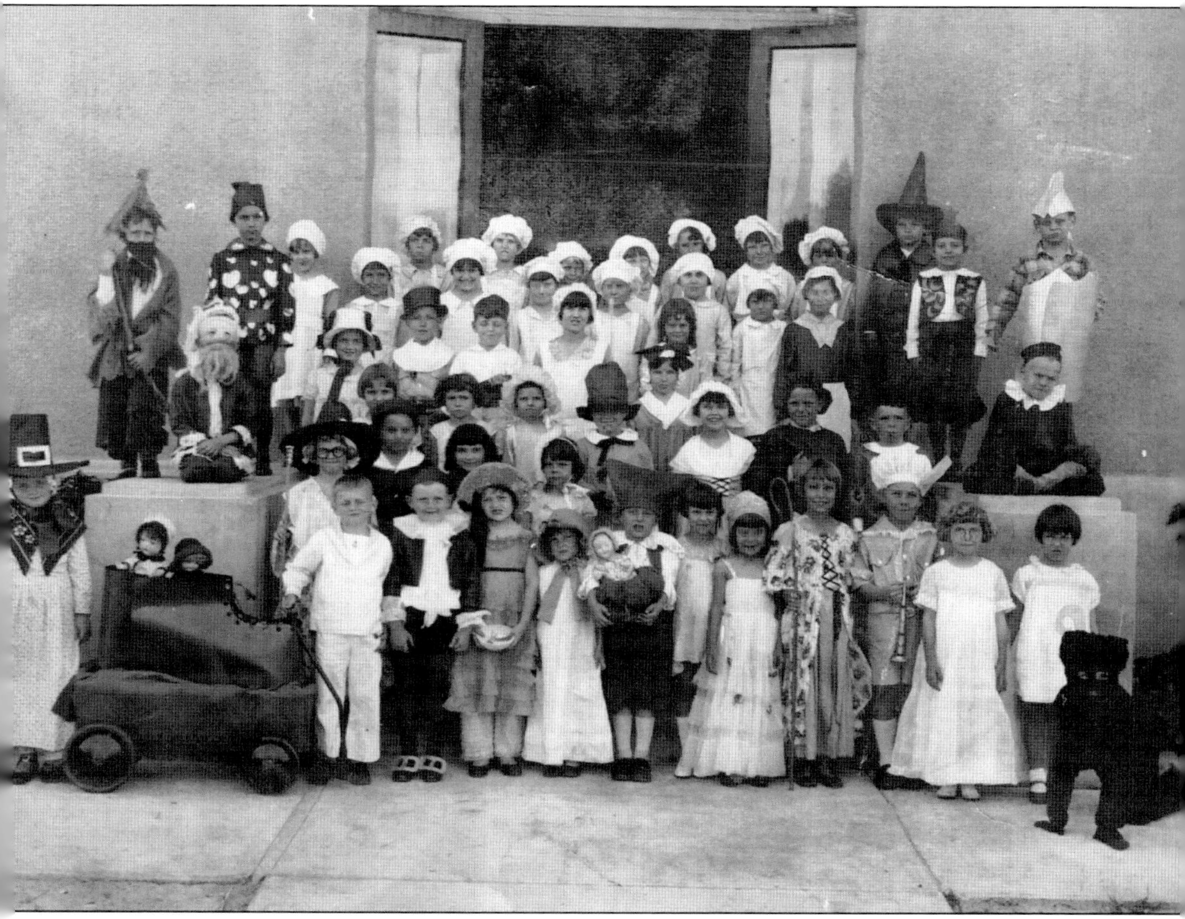

Palmetto Elementary Principal Frankie Howze was an advocate of school plays and would direct at least one annually. The children pictured in this 1930 Mother Goose play are, from left to right, (first row) Virginia Grimes, Buck McLean, George Harrison, Maxine Harrison, Mildred Henderson, Jimmy Howze, Elizabeth Brandon, unidentified, Jacqueline Kemp, Henry Smith, Catherine Buttram, and Dell Snyder; (second row) Louise Buttram, Tommy Ingram, unidentified, and Barbra Snyder; (third row) Lillie Curry, Jesse Waldref, Lenore Martin, unidentified, Edna Lackey, Martha Brandon, John Beville, and Sam Scott; (fourth row) Helen Grimes, Louis Menard, Alvin Rowe, Inez Wheeler, and Rosalie Parrish; (fifth row) Clara Hood, two unidentified, Bonnie Purvis, Lorene Thomas, unidentified, Mary Jo Taylor, and Josephine McBrayer; (sixth row) unidentified, Annie Lee Courtney, Lily Johnson, Mary Alice Barry, unidentified, Elizabeth Taylor, and Betty Colwell; (left pillar) Walter Wingert, Donald Grimes, and unidentified; (right pillar) Sam Bradley, Mary Louis Howze, Kenny Wooten, and Enoch Wingert. (Courtesy Lenore Stewart.)

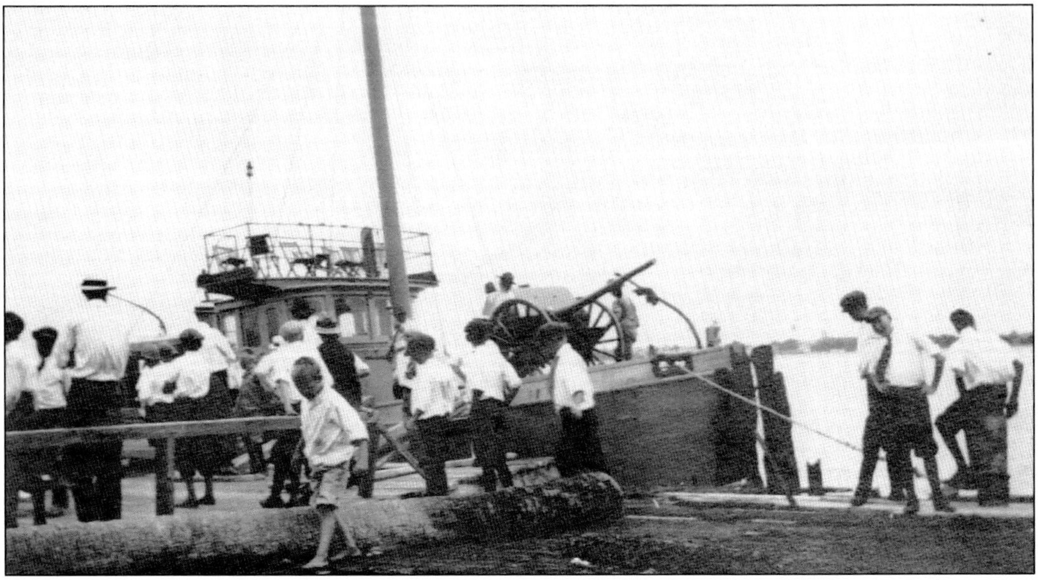

In this 1918 photograph, men unload a cannon from an Army gunboat for a victory festival marking the conclusion of World War I. During the firing of the cannon, the climax of the celebration, windows shattered in nearby homes from the concussion. (PHC.)

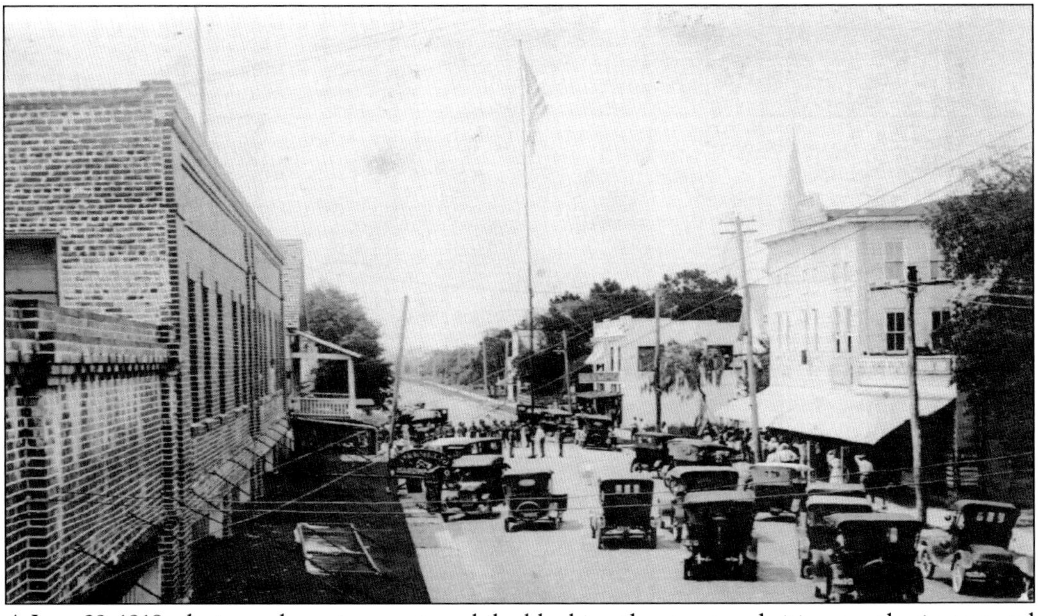

A June 29, 1918, photograph captures automobiles blocking the street and citizens gathering around to witness a new American flag raised for the first time at its home on Main Street. The Olympia Theater, on the right side of the street, was showing *The Spirit of Honor* at the time. (PHC.)

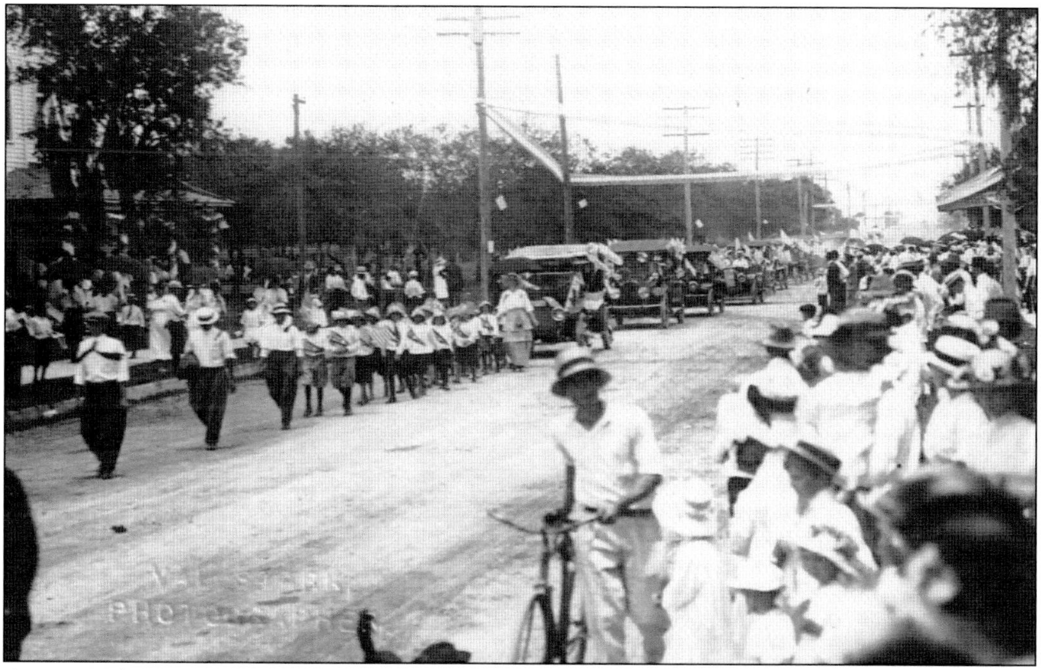

A seemingly ongoing procession of automobiles sets the scene for this c. 1920 Veteran's Day parade in Palmetto. Cheering residents are waving miniature flags in celebration. The size of the crowd gives a sense of Palmetto's population. (PHC.)

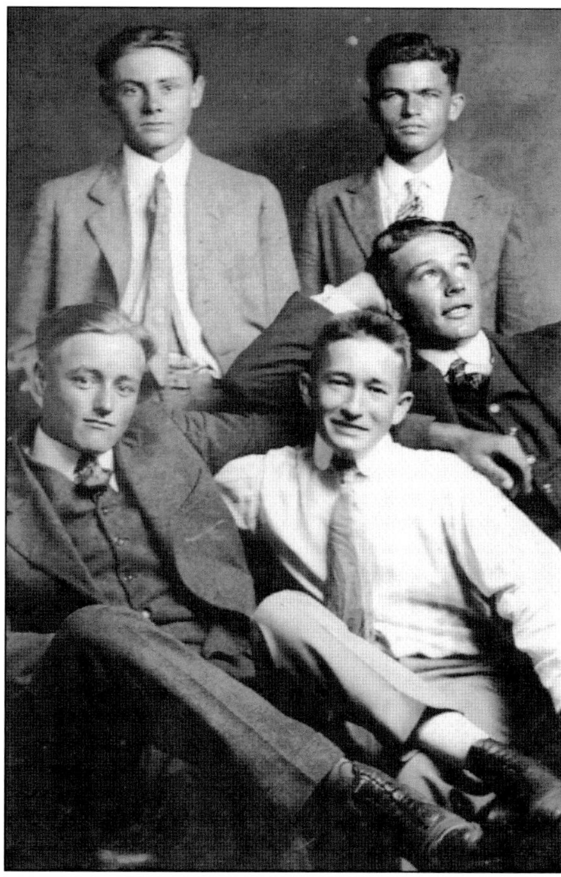

Pictured from left to right, Henry Hebble, William Toft, Paul Smith, and one of the Willis boys are dressed for a professional portrait taken at Colwell Photos. Palmetto girls found the young men heartthrobs in the 1920s. (PHC.)

• CLEANING and PRESSING

• ALTERING and REPAIRING

TAYLOR THE TAILOR

340 - 10th Avenue — Blue Building

PALMETTO

All Kinds of
GOOD HOMESITES FOR SALE . . . . EASY PRICES AND TERMS
In Several Locations By
$250 up — Pay $10 Month
TAYLOR the TAILOR (the owner)

Thomas F. Taylor, or "Taylor the Tailor," is considered Palmetto's most generous man. Originally from England, Taylor arrived in Palmetto in 1907. He was an entrepreneur, owning a music store, a movie theater, and a dry-cleaning and tailor shop. He vowed no child would ever pay more than a nickel to see a movie in his theater. Having none of his own, he often threw parties for children of all ages. He provided protection for kids through welfare agencies. The extent of his wealth remained unknown until his death in 1954, when his estate of more than $250,000 was donated to the Children's Homes of Florida. He set aside $500 for his burial, and when the bill was higher, the children of Palmetto returned his kindness by raising the money to pay for his headstone. Pictured below is a charcoal drawing of Taylor by Marlene Burton. (PHC.)

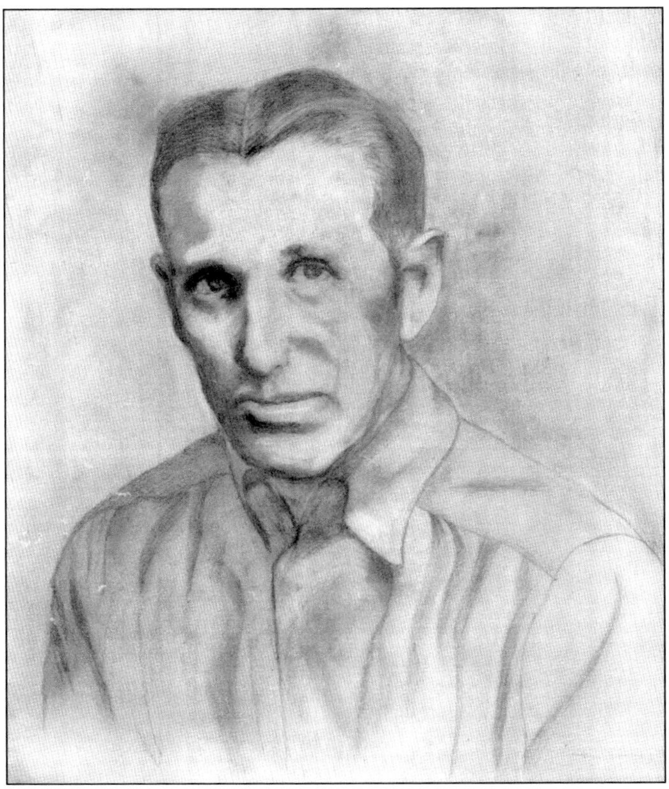

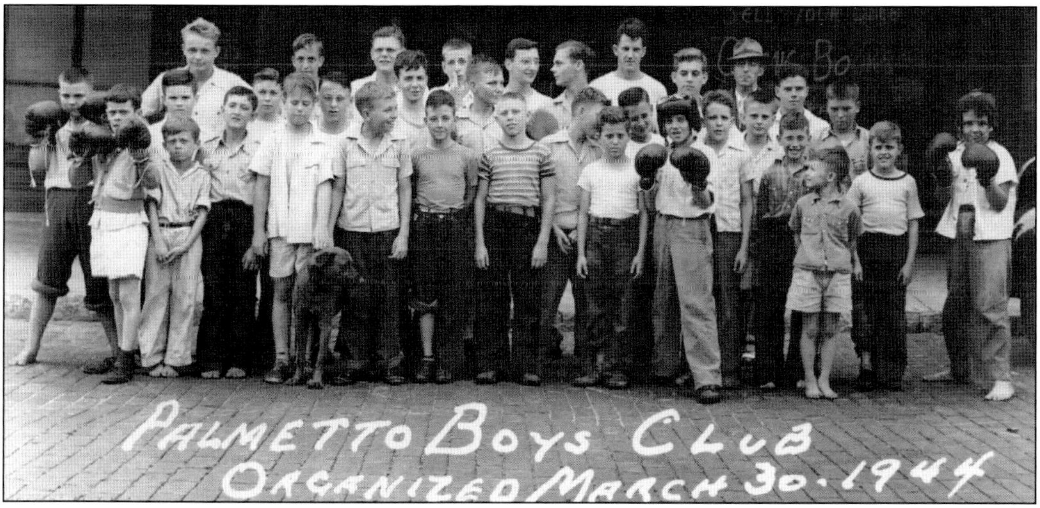

The Palmetto Boys Club was organized on March 30, 1944, although it was not officially part of a national network of partner organizations until 1957. A small wooden house near the present-day Boys and Girls Club on Tenth Street provided a haven for after-school activities for the children. Even after local schools were integrated in 1969, the club still barred African American youths until a few years later. (PHC.)

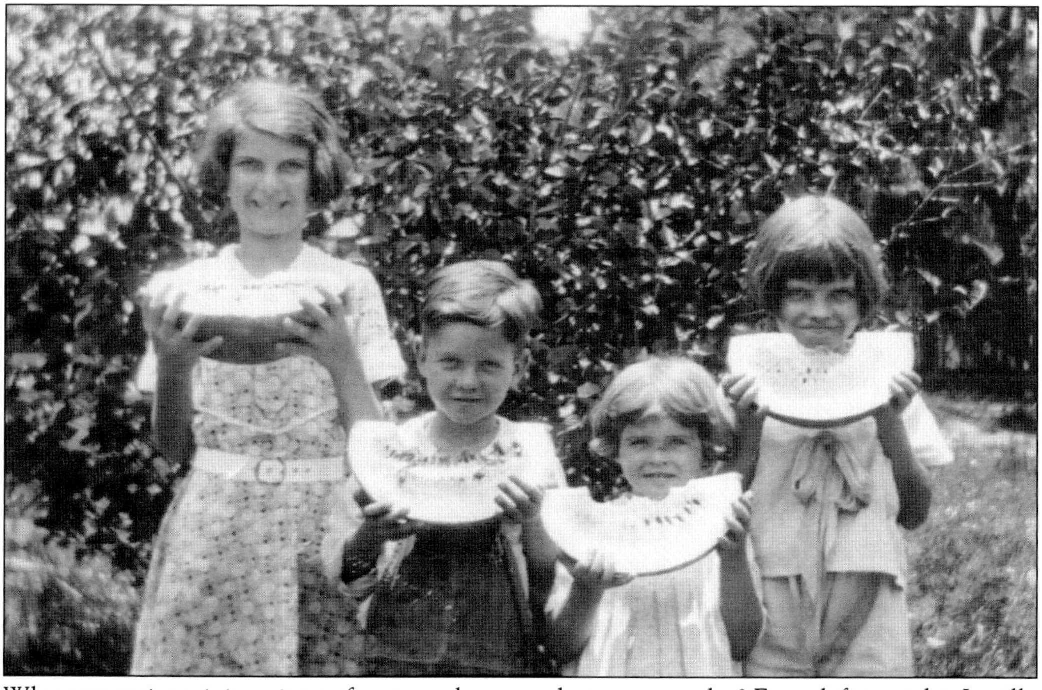

Who can resist a juicy piece of watermelon on a hot summer day? From left to right, Lucille Martin, Jimmie Martin, Jeanne Martin, and Lenore Martin waste no time devouring their treat in this 1930s photograph. (Courtesy Lenore Stewart.)

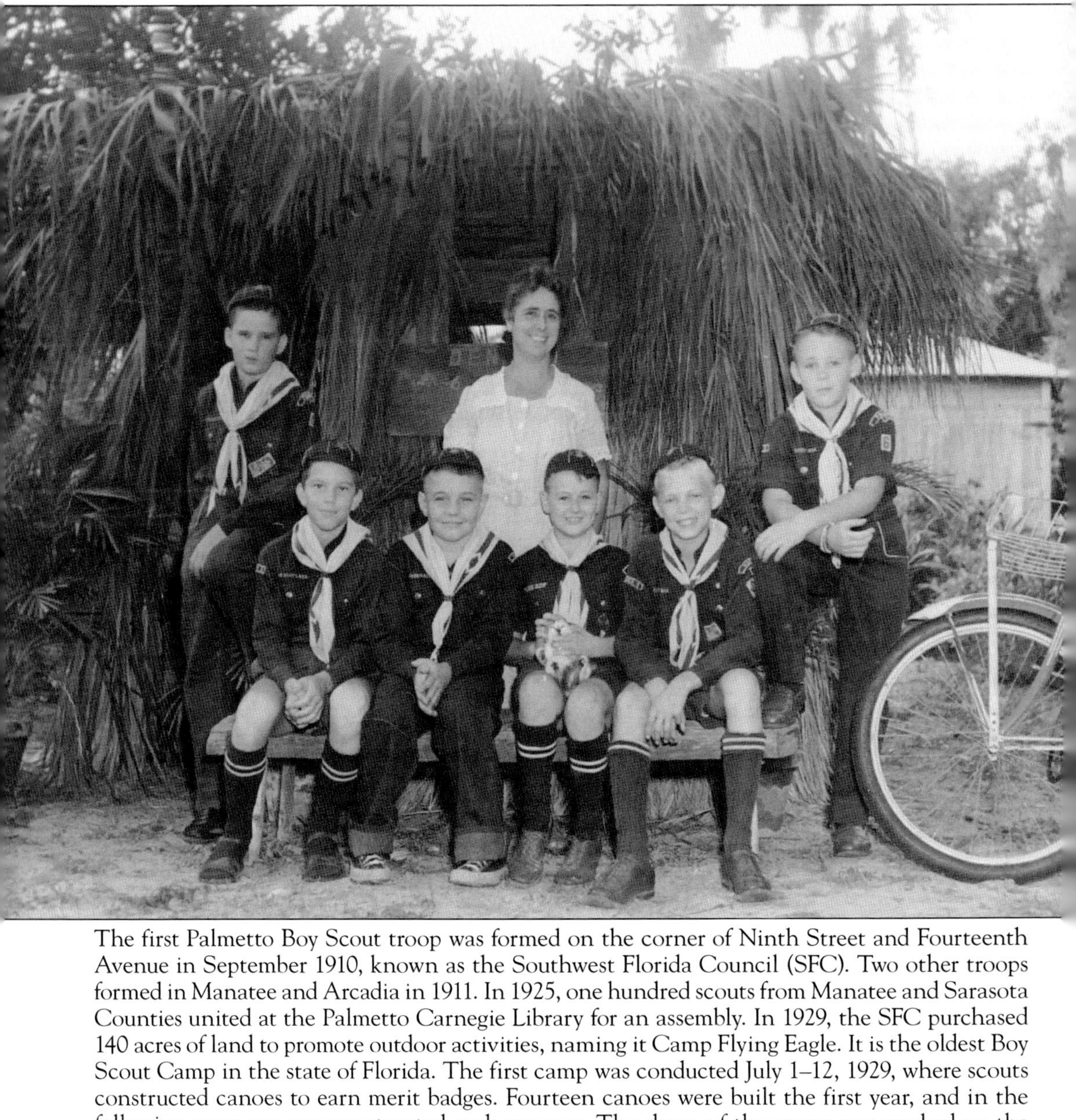

The first Palmetto Boy Scout troop was formed on the corner of Ninth Street and Fourteenth Avenue in September 1910, known as the Southwest Florida Council (SFC). Two other troops formed in Manatee and Arcadia in 1911. In 1925, one hundred scouts from Manatee and Sarasota Counties united at the Palmetto Carnegie Library for an assembly. In 1929, the SFC purchased 140 acres of land to promote outdoor activities, naming it Camp Flying Eagle. It is the oldest Boy Scout Camp in the state of Florida. The first camp was conducted July 1–12, 1929, where scouts constructed canoes to earn merit badges. Fourteen canoes were built the first year, and in the following years one was constructed each summer. The shape of the canoe was marked on the floor of the dining hall as a pattern; today, one can still see the faint outline. From left to right, David Haley, John Jackson, Louis Glayre, Elvira Glayre, David Long, Fellow Winters, and Mickey Harrison pose in front of a palm-frond fort. (PHC.)

Prior to 1948, two secondary schools served Manatee County's African American population: Lincoln High School in Bradenton and Memorial High School in Palmetto. Students studied in shabby buildings and used second-hand textbooks. The school board merged the two schools to stretch resources and increase the enrollment of Palmetto residents. When the new Lincoln Memorial High School opened in Memphis, students and teachers rejoiced. The high school became known throughout the state for its academic and professional achievements. Students were required to dress meticulously (right) and for once in their lives received decent textbooks (below). (Courtesy Lincoln Middle School.)

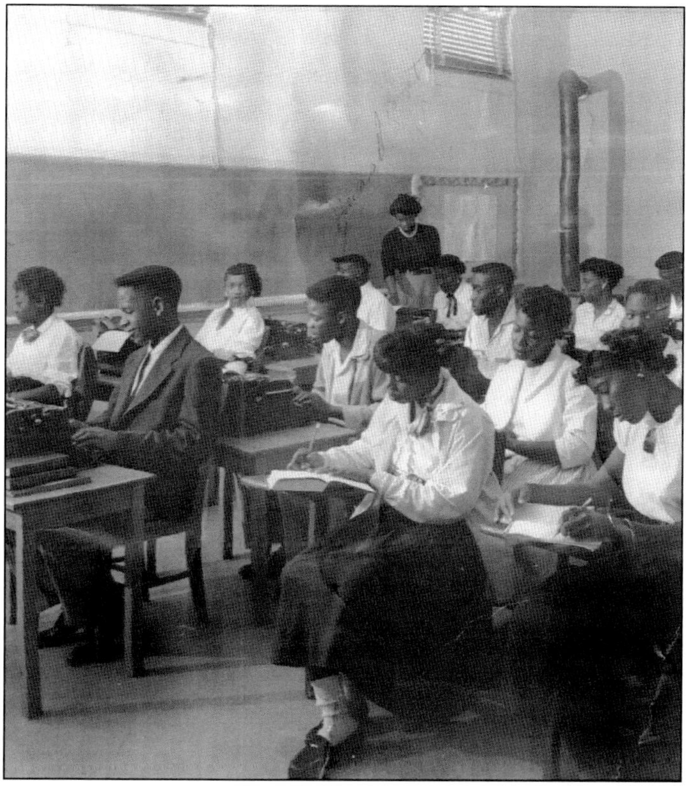

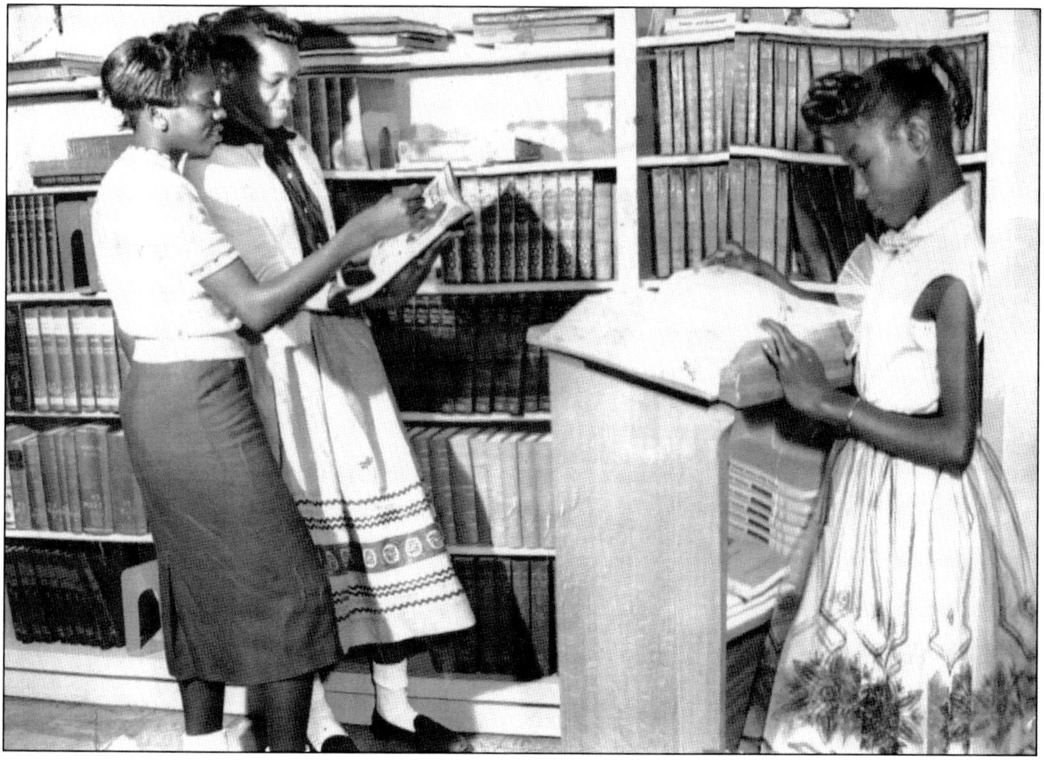

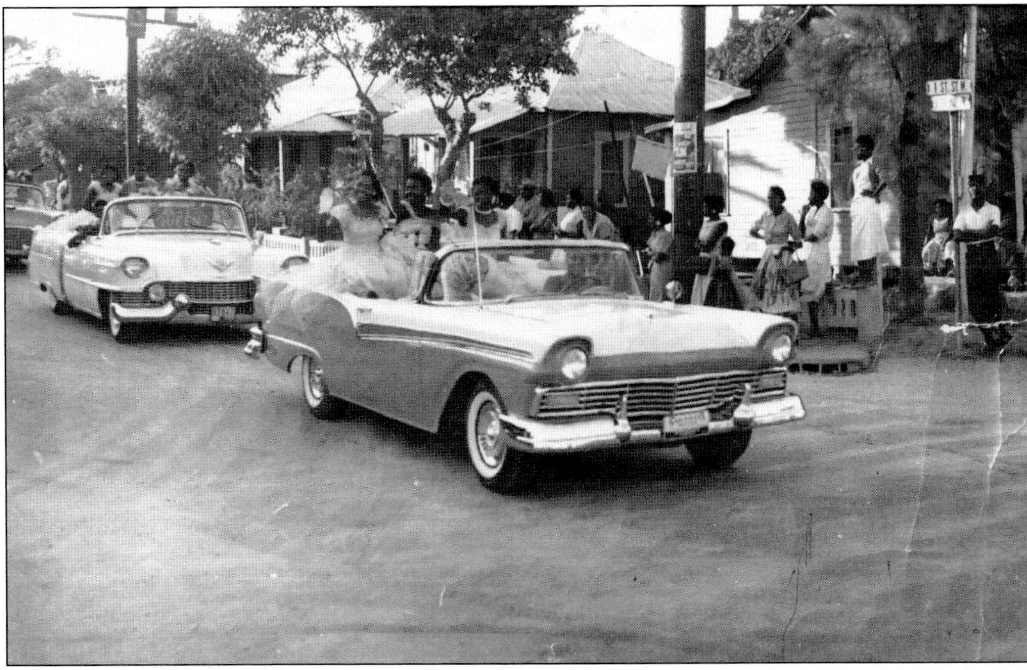

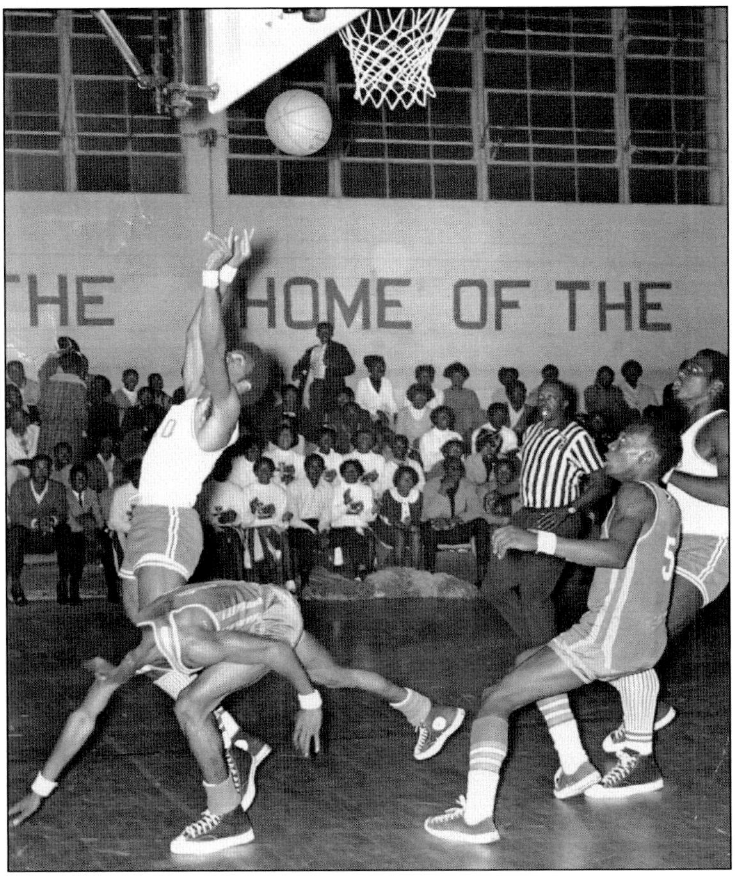

Lincoln, accredited by the Southern Association of Colleges and Schools, was the pride of the black community and earned its unflagging support. Here, people flock to catch a glimpse of the homecoming court parading down Seventeenth Street East. (Courtesy Lincoln Middle School.)

Every Friday night, residents from all over the county would come out to watch the ball game. The stadium seats are completely full in the background of this action shot of the Lincoln Trojan basketball team. (Courtesy Lincoln Middle School.)

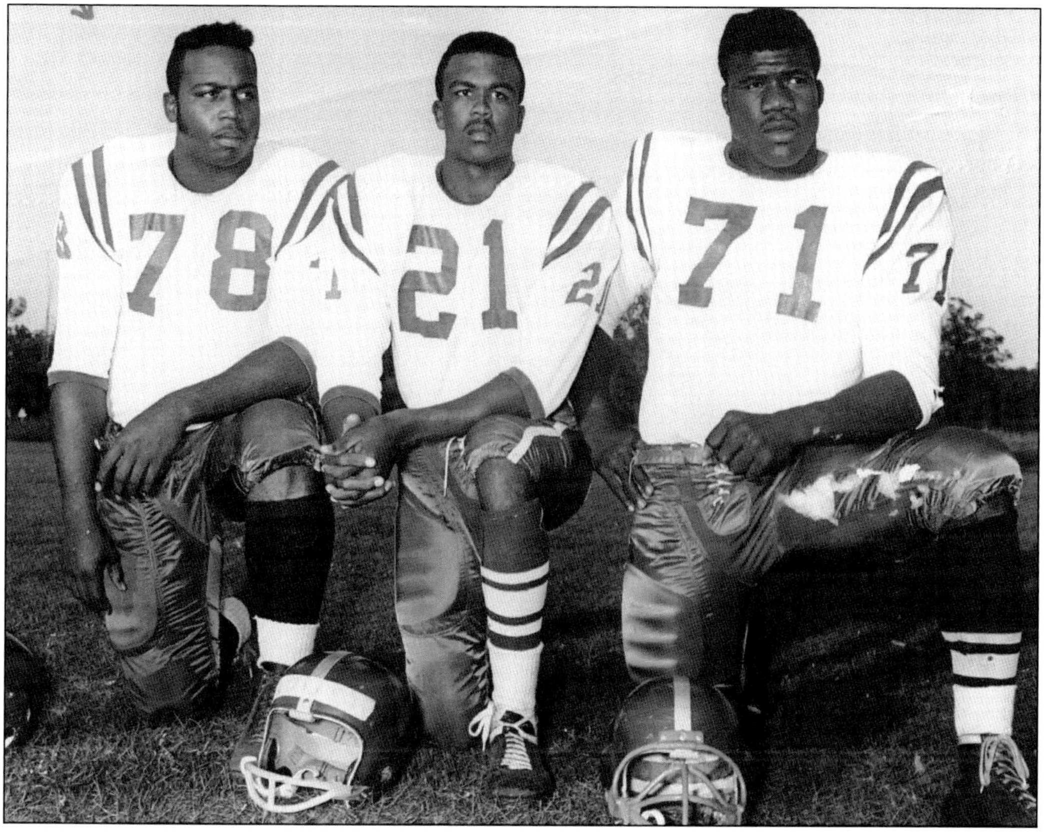

Both Memorial and Lincoln High Schools had their own sports teams that were coached by volunteer teachers and community members. Upon their consolidation in 1948, the student-body and faculty representatives selected the Trojan as the school mascot. The prowess of the Lincoln Trojan football team became renowned throughout Florida. US 41 separated the playing field from school grounds. A tunnel dug beneath the highway ensured safe passage for the players. Traveling "through the tunnel" before a game became a beloved ritual for the players, allowing them to pass symbolically from a world of injustice and emerge as respected athletes. The ritual inspired the award-winning documentary *Through the Tunnel*, directed by Durand Adams, Charles E. Williams, and Charles Clapsaddle. The tunnel still joins Lincoln Park and Lincoln Middle School. (PHC.)

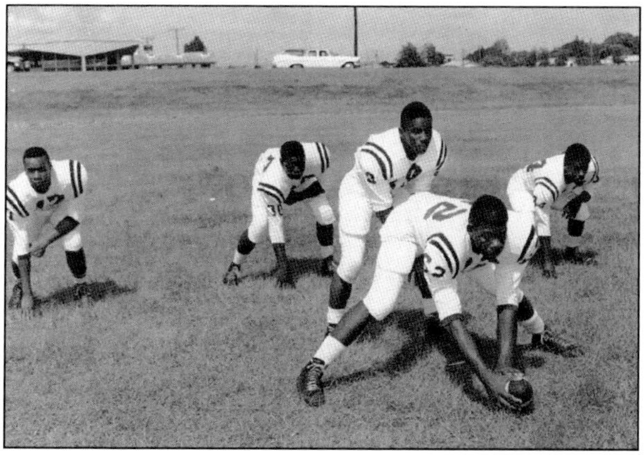

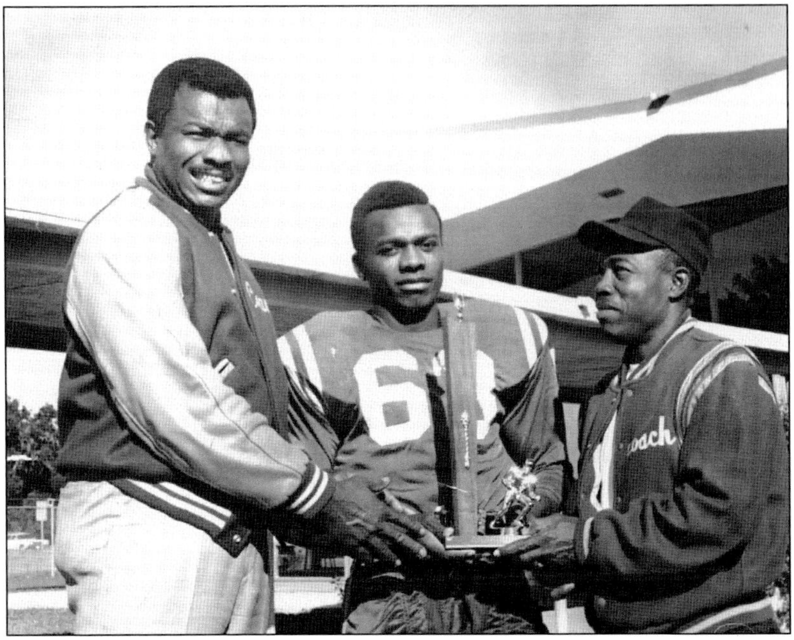

Lincoln's head coach Eddie Shannon (far right) encouraged athletes to follow their dreams. He was a 54-year coaching veteran and an integral figure when Lincoln was integrated in 1969. He pushed the integrated football team to set an example for white and black students and their families, an undertaking that bonded the student body. (PHC.)

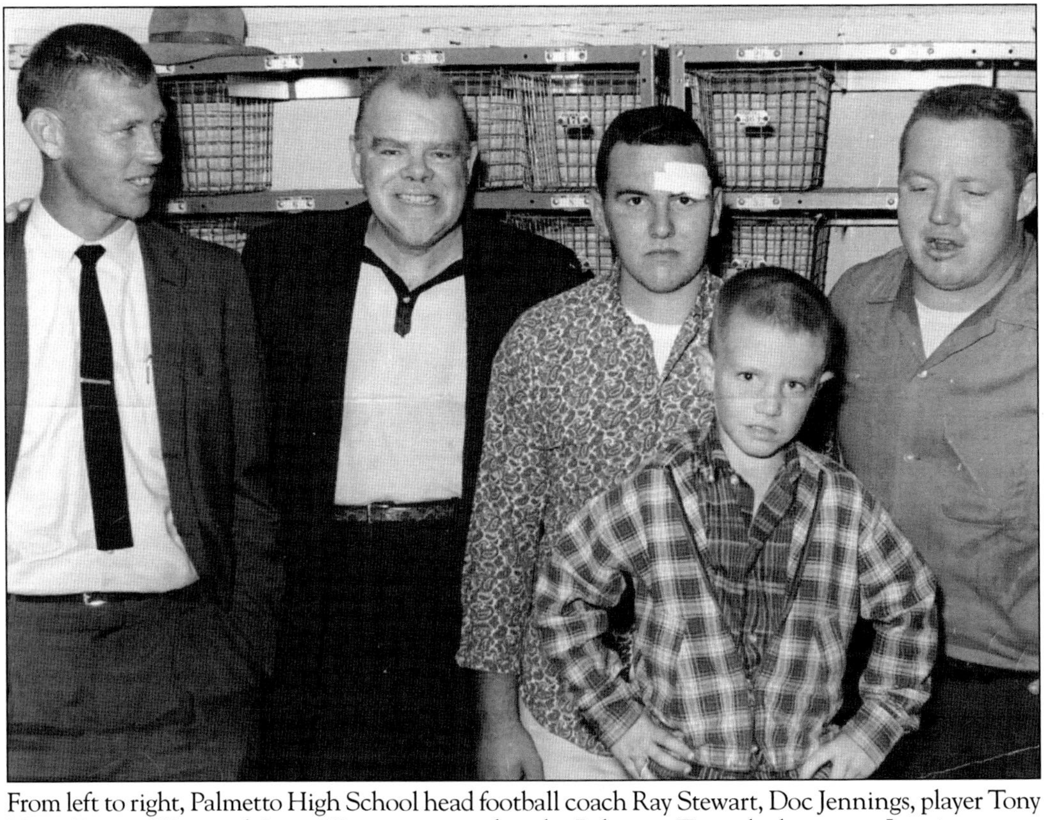

From left to right, Palmetto High School head football coach Ray Stewart, Doc Jennings, player Tony Haug, Ronny Witt, and Gene Witt are pictured in the Palmetto Tigers locker room. Jennings was a local physician who traveled to all the games, offering his services to injured players. At this game, Haug's forehead was split open and had to be stitched up at halftime. (Courtesy Tony Haug.)

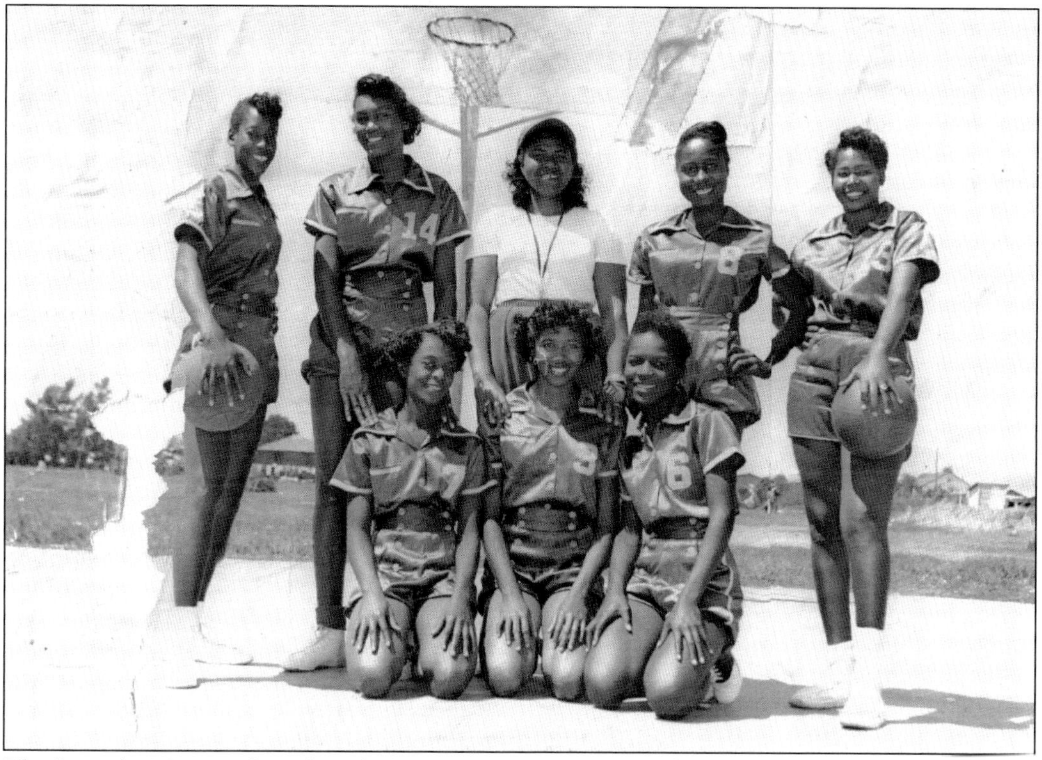

The Lincoln Memorial High girls' basketball team poses with coach Mildred Dawson in new uniforms. Until the new Lincoln school opened in 1948, players' uniforms were often hand-me-downs donated from white schools. (Courtesy Lincoln Middle School.)

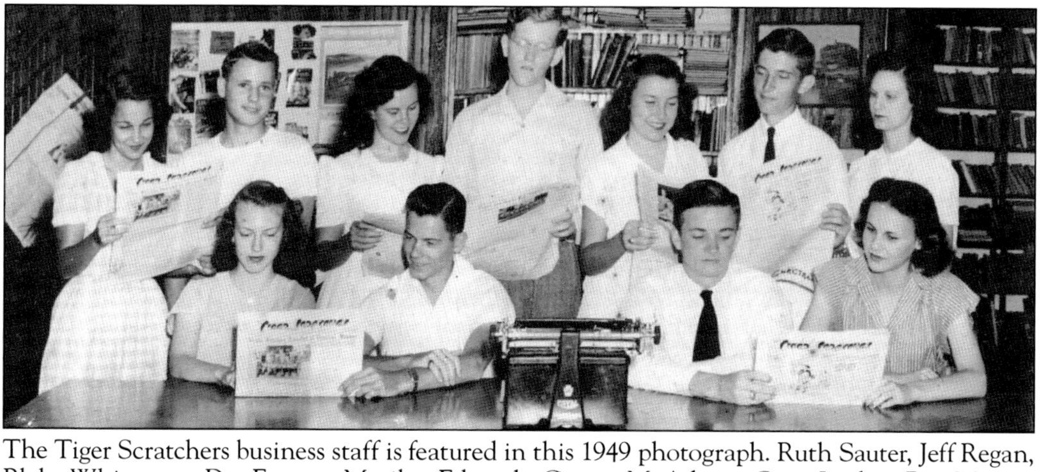

The Tiger Scratchers business staff is featured in this 1949 photograph. Ruth Sauter, Jeff Regan, Blake Whisenant, Dot Everett, Marilyn Edwards, George McAdams, Gene Lackey, Ben Mount, Lavonne Candle, Ben Tillett, and Lynette Hines are reading copies of the newspaper they produced with the manual typewriter seen in the foreground. (PHC.)

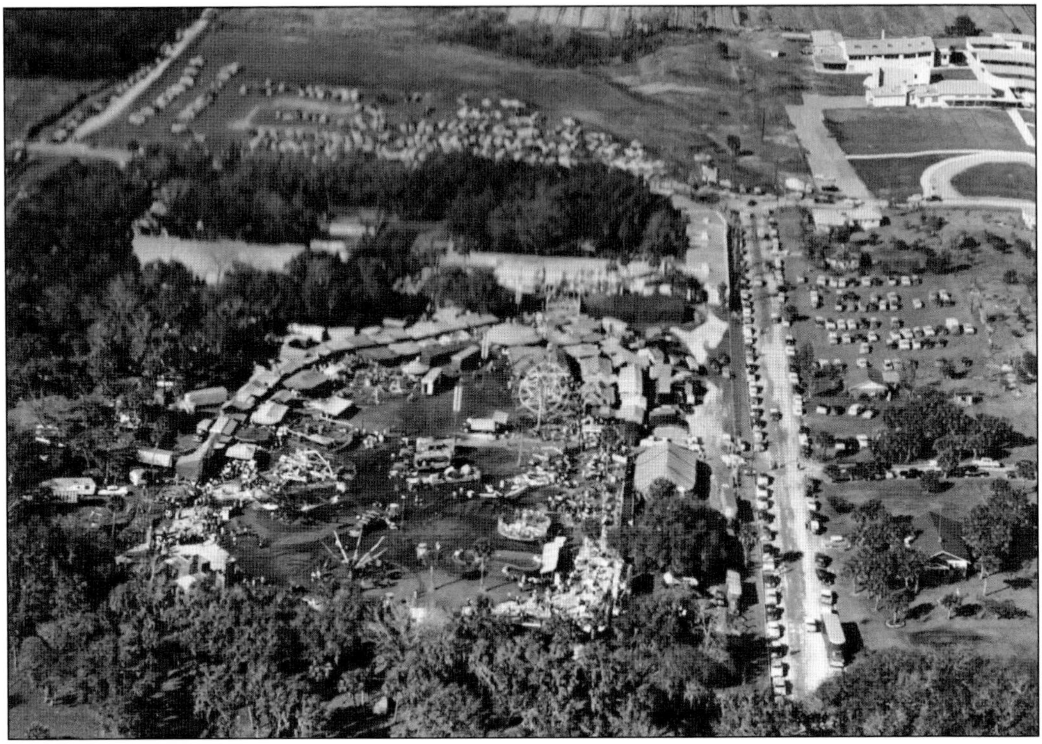

The first fair was held on February 28, 1916, on Ninth Street in Braidentown. The board of trade hired contractor E.W. Stoltz of Palmetto to begin work on the main exhibition building. A grandstand was constructed to view fair entertainment. The buildings were erected on the present McKechnie Field site. In the early 1930s, during the Great Depression, Bradenton stopped sponsorship of the fair, but before long, 35 Palmetto men signed application for a charter. In 1937, the fair was held in a corrugated-metal building near the Palmetto Trailer Park. Exhibits were held in January 1942, but the fair was then discontinued during the war years before resuming in 1951. An aerial view of the 1958 fair in Palmetto is pictured above. Below, men are smoking mullet outside the metal building in Palmetto. (PHC.)

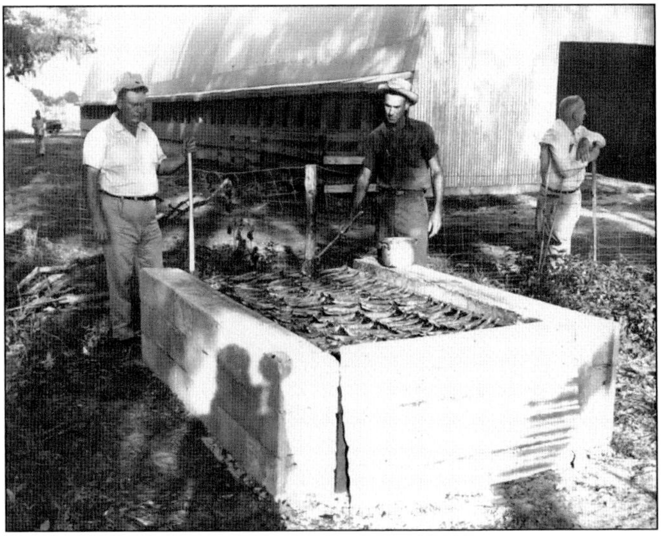

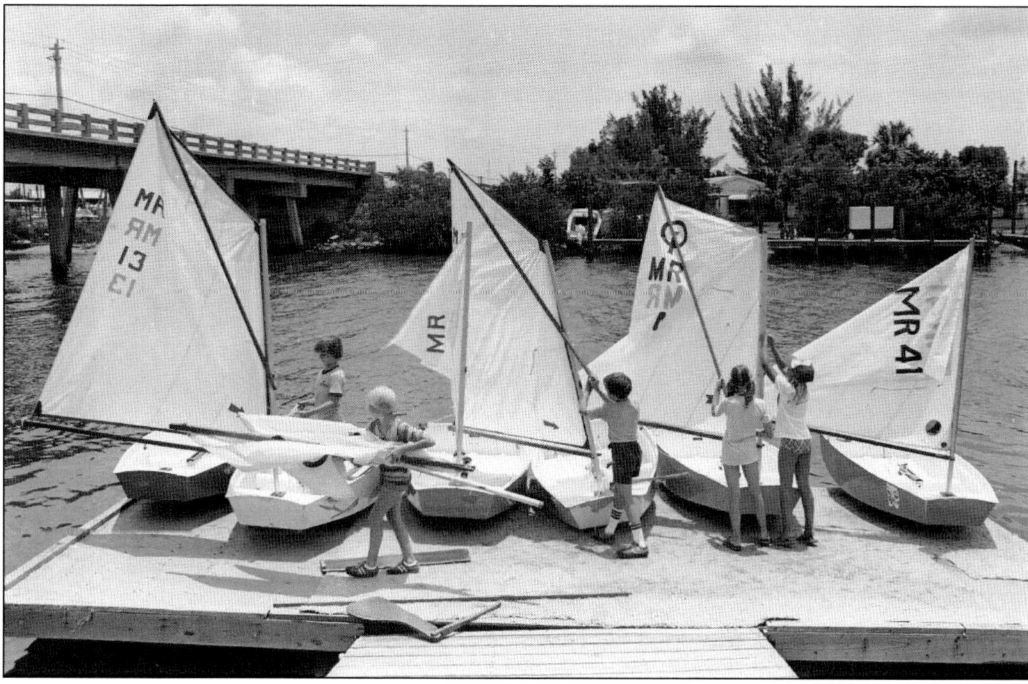

Clearwater resident Clark Mills invented the optimist pram in 1947. He built the original out of a sheet of plywood. The boat was an overnight success, as locals lined up to buy plans that produced a sailboat for under $50. The Manatee River Boat Club formed the Manatee River Pram Fleet a short time later. Harold "Pappy" Stoehrer built three boats to start the organization. Local merchants sponsored the cost of the prams. On October 5, 1953, the pram shelter opened to much fanfare. It was originally located at Palmetto Pier (above). Fifty-six prams participated in the dedication regatta, which at the time was the largest sailing race ever held in Florida. By 1955, the Manatee River Pram Fleet was the largest fleet of single-design boats in the state. In 1958, the pram shelter was moved to its current location on Snead Island (below). (MCHS.)

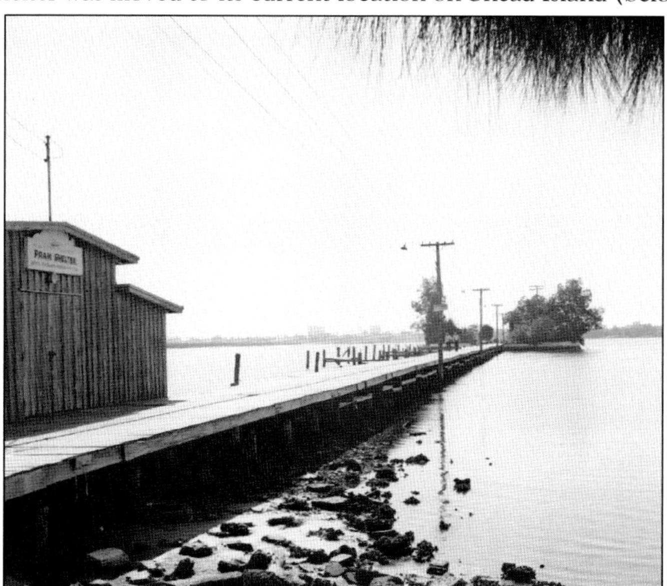

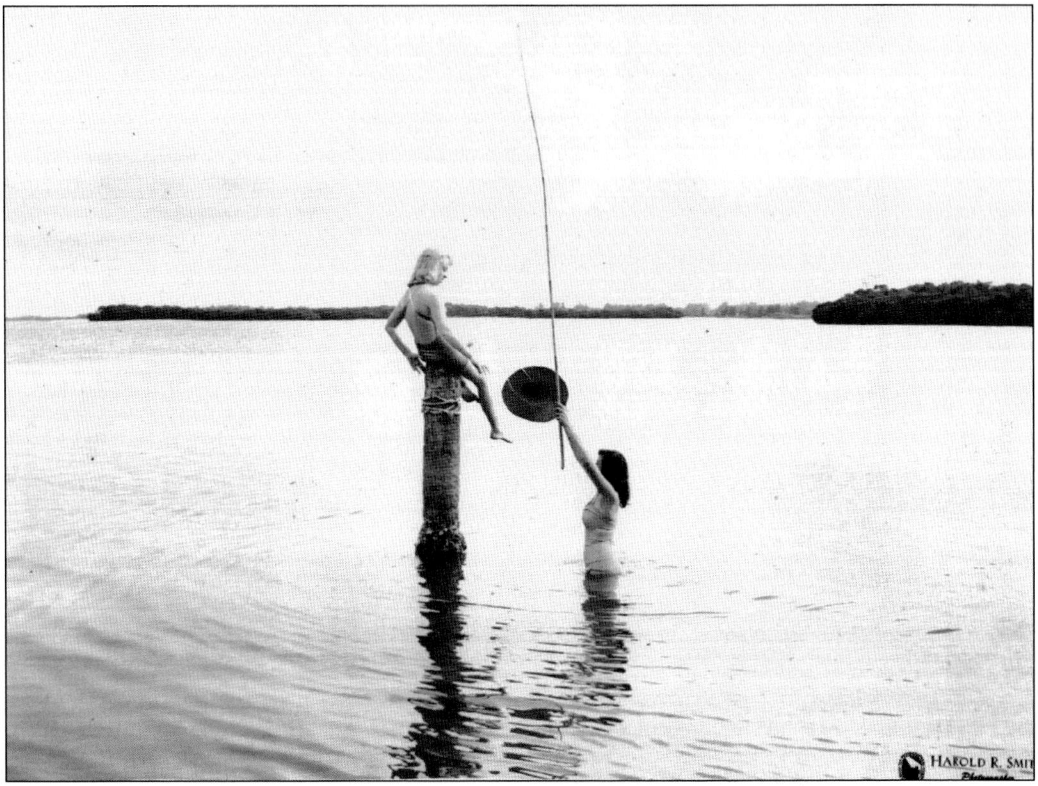

Two young ladies pose for a picture in Terra Ceia Bay. The Palmetto Chamber of Commerce hired photographer Harold R. Smith to take a series of such photographs for city publicity purposes. The style of swimsuits places the date sometime in the 1940s or 1950s. (PHC.)

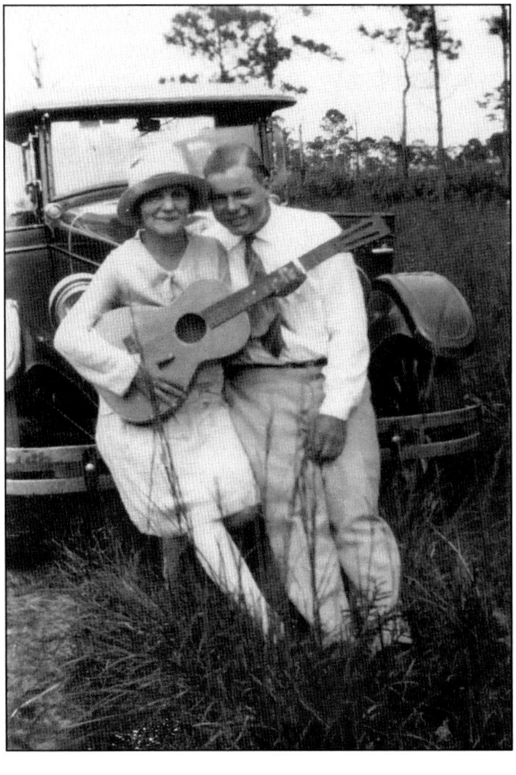

A happy couple has stopped after a drive into the countryside for a picnic lunch—a daylight serenade beneath the pines was sure to follow. (SLAF.)

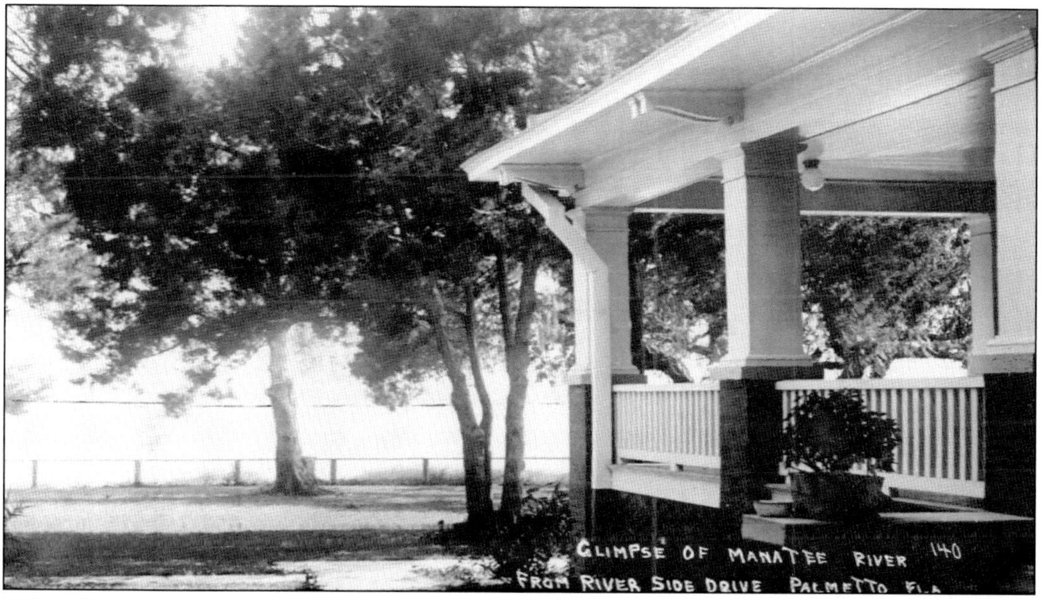

Life in Palmetto meant adapting to the natural surroundings. Houses often incorporated outside scenery into the structural design. Residents of this house most likely use free time sitting in a rocking chair on the porch, glimpsing out at the Manatee River. (PHC.)

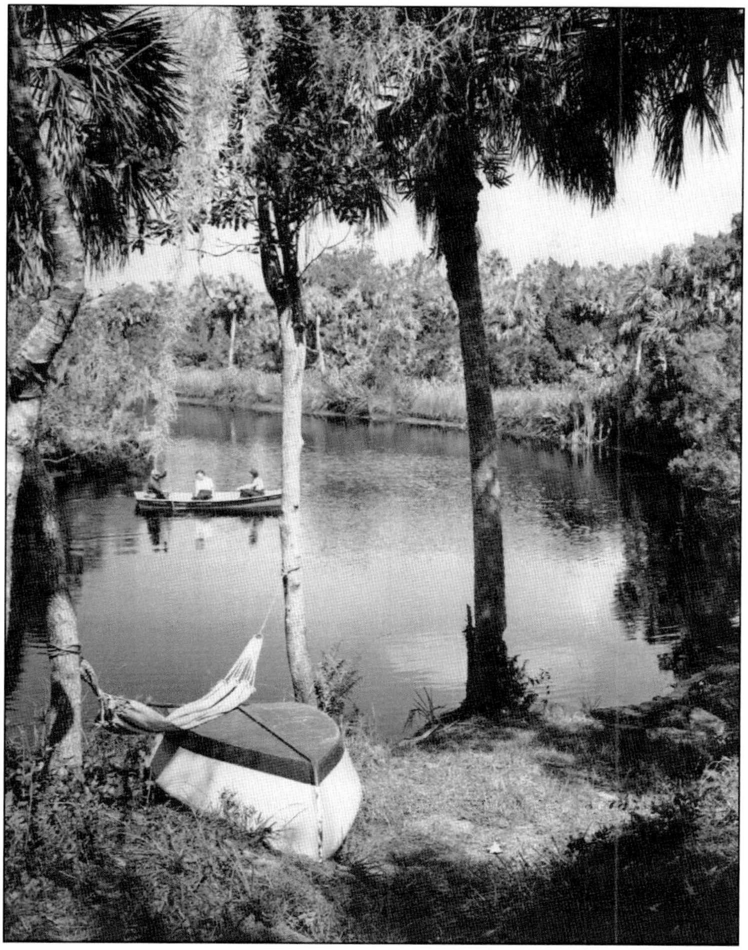

Photographer Harold Smith captured this serene setting on the upper Manatee River. The lovely Florida fauna is reflected in the dark waters. An unoccupied hammock hangs between two trees, beckoning the boaters to take a rest when they get back to shore. It is another perfect day. (PHC.)

Discover Thousands of Local History Books Featuring Millions of Vintage Images

Arcadia Publishing, the leading local history publisher in the United States, is committed to making history accessible and meaningful through publishing books that celebrate and preserve the heritage of America's people and places.

Find more books like this at
www.arcadiapublishing.com

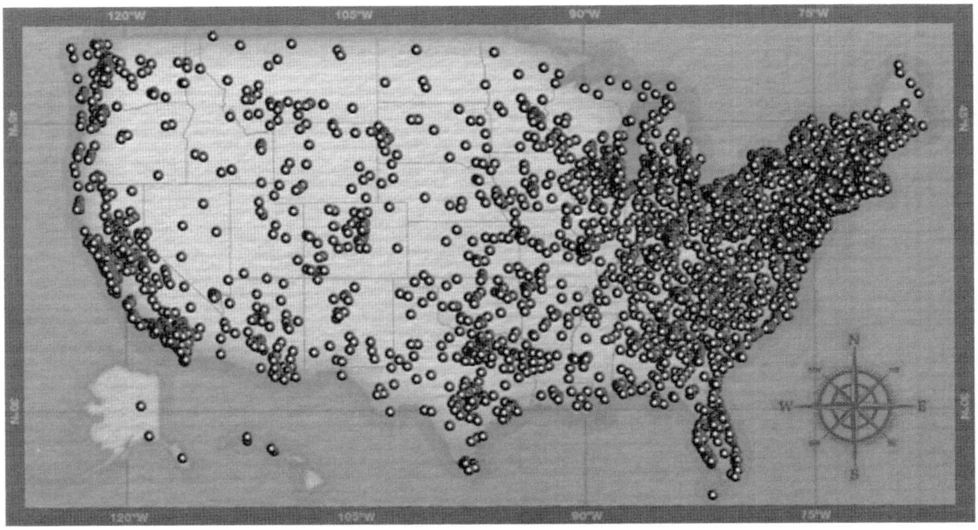

Search for your hometown history, your old stomping grounds, and even your favorite sports team.

Consistent with our mission to preserve history on a local level, this book was printed in South Carolina on American-made paper and manufactured entirely in the United States. Products carrying the accredited Forest Stewardship Council (FSC) label are printed on 100 percent FSC-certified paper.

MADE IN THE USA